for Catherine

Object*ivity*

from David

Surveyor's theodolite base
Three sets of double legs attach to the six sockets, forming the stand to which the theodolite is fixed.
21 cm/8¼ in.

Object*ivity*

A designer's book of curious tools David Usborne
Foreword by Thomas Heatherwick

Acknowledgments

My principal debt is to Pearce Marchbank who first suggested this book and who, with assistance from Ben May and with photography by Celine Marchbank, has given it this handsome form.

The staff at Thames & Hudson have with patience and skill helped us transform what began as a collection of photographs into a publishable volume.

Over the years that it has taken to finish this project I have been grateful to three friends, Emmanuel Cooper, Michael Glickman and David Batterham, for their advice about the world of publishing.

Fellow collectors and dealers have been generous with time and information. I would like to thank Michael Graham-Stewart, Anthony Jack, Willy Stewart, David Burns, Maurice Collins, Alison Milner, Vivian and Gretchen Anderson, John and Valerie Arieta.

Even though they may not like what I like, it helped when visiting street markets or antique fairs to have friends along. For their company, tolerance of my obsessions and inspired donations to my collection I thank Tess and Julyan Wickham, Peter and Glenys Major, Rita and Murray Scher, Ilse and David Gray.

Nobody learns to see without tutors and guides. Mine are, or were, Ian Fleming-Williams, Hans Schleger, Steen Eiler Rasmussen, Aaron Suskind, Michael Wickham, Aldo van Eyck and Frank Maresca.

I owe special thanks to Leila McAllister without whom I would not have met Thomas Heatherwick and Maisie Rowe and through them my publishers.

For help in identifying and describing medical instruments I am grateful to Robert McGibbon and Ronnie D'Silva.

Tony Murland, Roy Arnold, John Collier and Steve White all made useful identifications of puzzling woodworking and metalworking tools.

Gilbert Lewis reassured me about my anthropological speculations and Dawn Ades was similarly supportive in the complex area of Surrealism, though neither necessarily endorses my views.

Finally, this book is dedicated to the memory of James Alfred Brew, 1923–2008.

David Usborne

Design by Pearce Marchbank RDI
Layout by Ben May
Photography by Celine Marchbank

First published in the United Kingdom in 2010 by
Thames & Hudson Ltd, 181A High Holborn, London WC1V 7QX

thamesandhudson.com

Copyright © 2010 David Usborne
Foreword copyright © 2010 Thomas Heatherwick

All Rights Reserved. No part of this publication may be reproduced or transmitted in any form or by any means, electronic or mechanical, including photocopy, recording or any other information storage and retrieval system, without prior permission in writing from the publisher.

British Library Cataloguing-in-Publication Data
A catalogue record for this book is available from the British Library

ISBN 978-0-500-51501-3

Printed and bound in China by Hong Kong Graphics & Printing Ltd.

Motorcycle cylinder head
The metal fins help to cool the engine.
20 cm/7³⁄₄ in.

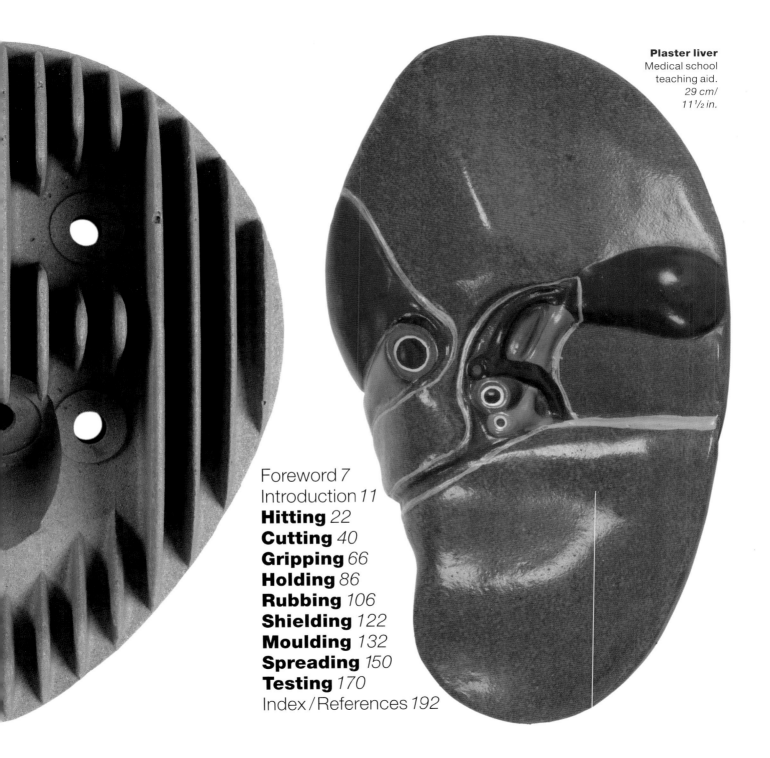

Plaster liver
Medical school teaching aid.
*29 cm/
11½ in.*

Foreword *7*
Introduction *11*
Hitting *22*
Cutting *40*
Gripping *66*
Holding *86*
Rubbing *106*
Shielding *122*
Moulding *132*
Spreading *150*
Testing *170*
Index / References *192*

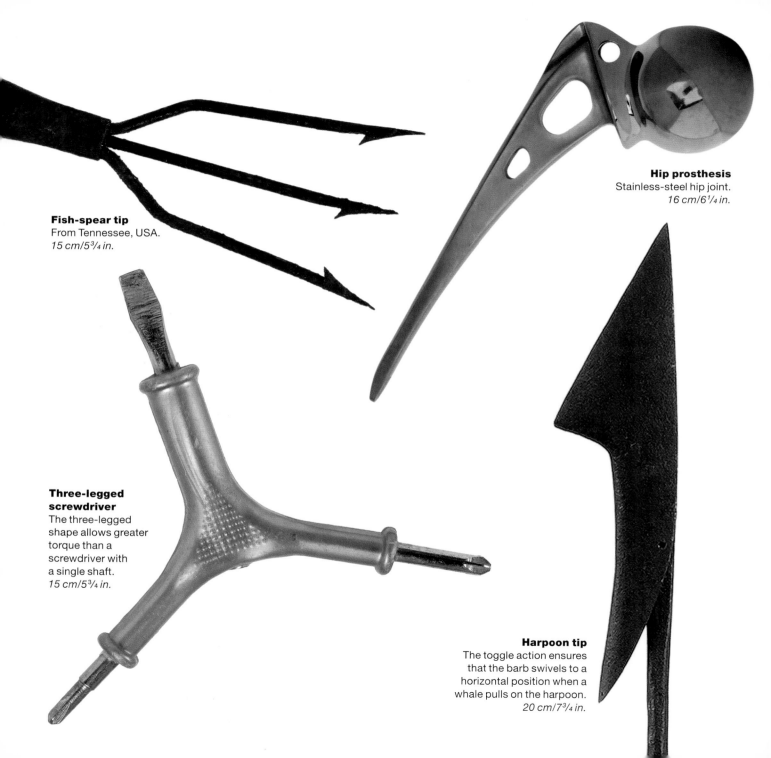

Fish-spear tip
From Tennessee, USA.
15 cm/5¾ in.

Hip prosthesis
Stainless-steel hip joint.
16 cm/6¼ in.

Three-legged screwdriver
The three-legged shape allows greater torque than a screwdriver with a single shaft.
15 cm/5¾ in.

Harpoon tip
The toggle action ensures that the barb swivels to a horizontal position when a whale pulls on the harpoon.
20 cm/7¾ in.

Foreword

I first encountered this extraordinary collection after curating an exhibition for the Conran Foundation at London's Design Museum in 2004. I had agreed to collect £30,000-worth of items representing 'good design', but found myself more interested in the idea behind each product than by the product itself. This became then a project to find and exhibit one thousand ideas, each of which in my opinion embodied some kind of ingenuity – regardless of how functional, good-looking or desirable the product was.

The project took me to Beijing hardware stores, Berlin sex shops, Polish builders' merchants and an Istanbul kebab shop equipment supplier, and it brought me into dialogue with the inventor of a plastic bag for carrying pizzas, a goat farmer, a supplier of mortuary equipment and a Glasgow sausage-skin manufacturer.

David Usborne saw the resulting exhibition and invited me to see this collection. The visual richness of the objects instantly appealed to me as a designer experimenting with ideas, shapes, materials and ways of making things. While most of the objects are functional tools, rather than products of an artistic imagination, they are also extraordinarily expressive, suggestive of faces, figures, animals or movements. It is easy to imagine these objects coming alive at night to roll, spring, beat, ping, scratch, pump or slice when the humans sleep.

It struck me too that this is a nostalgic collection, that these are the orphans, misfits and rejects of industrial change and human development. For many of the objects, the point is not that they have been replaced by new technology, but that they do something that no longer needs to be done. Like the tools associated with crafts which are no longer practised, such objects have something to say about past lives.

Every single object in this collection is gorgeous and I confess to coveting them all.
Thomas Heatherwick

Two hoes
Hoes are miniature ploughs attached to handles. The larger hoe has classic leaf-shaped blades like that on the right of page 35.
Right: 20 cm/7¾ in. Far right: 26 cm/10¼ in.

'The real connoisseurs in art are those who make people accept as beautiful something everybody used to consider ugly, by revealing and resuscitating the beauty in it. Those are the only true connoisseurs: the rest are the blind slaves of the prevailing fashion.'

Goncourt Journal, *Paris, 1881*

Tyre iron
This mundane garage implement exactly follows the shape of William Hogarth's 'line of beauty'.
35 cm/13¾ in.

Wooden pillow
African headrest resembling an executioner's block.
Hungry rats have left tooth marks on the curved edge.
22 cm/8¾ in.

Gas connector
A flexible metal pipe connecting the gas supply to a water heater.
15 cm/5¾ in.

Introduction by David Usborne

The desire to collect takes many forms, which wealth, opportunity and taste determine. While the rich have the means to collect houses, horses, islands and yachts, it is not necessary to have money to satisfy the collecting urge. Prisoners in solitary confinement have kept their spirits alive by organizing miniature zoos of the insects that venture into their cells. Collectors of ephemera hover outside supermarkets scavenging for discarded shopping lists while others loiter beside photobooths eager for rejected self-portraits.

Collecting has been described as a form of madness, but it can also be a way of surviving situations of loneliness and deprivation. Sent away to school at the age of eight, I found myself surrounded by schoolboy collectors of everything from conkers, sweet papers and stamps to cigarette cards, butterflies and comics. Lonely, in a situation in which we had few choices, our collections were little worlds that we could control, in which we tentatively expressed our emerging personalities and through which, as we swapped and bartered, we learned about each other and the rudiments of capitalism.

If it is a form of madness, it is interesting to note how many creative and apparently sane artists have suffered from it. Vladimir Nabokov collected butterflies, Edgar Degas paintings, Lord Byron, it is said, his lovers' pubic hair. The largest room in Rembrandt's house was filled with the strange objects he used as props in his portraits. Sigmund Freud, when he sat at his desk, faced a small museum of Greek, Roman and Egyptian figures, some of which were occasionally chosen to stand opposite him as he dined.

Because artists' collections can shed light on the sources of their art, I have always found domestic photographs of painters and sculptors fascinating. My eye swerves past Pablo Picasso or Georges Braque to the African and Oceanic figures on their shelves; past Jackson Pollock and Lee Krasner to the broken ship's anchor on their wall; past Henry Moore to his working collection of rocks, roots and bones; past Constantin Brancusi to the strange, bifurcated stove that heated his studio and the huge threaded wooden column from a wine press that lurked in the shadows.

Nobody should be ashamed of collecting, nor should those who have never felt the urge feel superior. In the closing shots of Orson Welles's *Citizen Kane* the camera glides over the accumulated relics of Kane's life. The film seems to sneer at the vast array of expensive art and assorted junk with which a rich man has tried to cheat death. Yet it is through these objects, the young newspaper owner's Declaration of Principles, the shattered globe or the boy's sledge, blistering in the furnace, that Kane's story is told.

If their collections survive, collectors do, in some sense, cheat death. Without them the world's great galleries and libraries would not exist; there would be no British, no Metropolitan, no Smithsonian museums. However, as these institutions have grown, as they have become collections of collections, they have inevitably lost some of the excitement of their origins. At the British Museum, Hans Sloane's cabinet of curiosities is now buried under a great coral reef of objects acquired from other collectors. Aware of this, and in an attempt to give the visitor something of the true flavour of collecting in the 18th and 19th centuries, the museum has recently filled the cavernous space of the former King's Library with a deliberately jumbled array of treasure from its storerooms. The same excitement can be felt, a short walk away, in the mirrored congestion of Sir John Soane's Museum in Lincoln's Inn Fields.

One of the advantages of individual collections is that the owner, as the curator, is free to organize their possessions according to any system of classification they choose. At the Pitt Rivers Museum in Oxford the artefacts of different tribal societies are grouped by function rather than geographical origin. The visitor can compare clubs from six different islands in the Pacific and then pass on to a similarly varied display of jewelry or textiles. Other collectors, such as André Breton and Sir John Soane, preferred to arrange their possessions more for aesthetic effect than academic instruction. They enjoyed strange juxtapositions and the chance encounters produced by deliberate disorder.

What kind of collection is recorded in this book? What is the thread that connects a metal eel spear and a wooden hat block to a ceramic hot-water bottle? What caught the collector's eye at the antique fair, flea market or junk yard from which any of them might have come?

At first sight they look like a selection of goods from the shelves of an eccentric hardware shop. There are tools here for plumbers, farmers, gardeners, surgeons, cooks, sculptors, hatters, roofers, masons, butchers and builders. Yet, despite the variety of materials and sources, each item satisfies all of their collector's criteria.

They are, or were, useful tools, in the widest sense of the word, which covers all technology. Their function is often difficult to guess, unless you belong to one of the specialized trades listed above. They have almost no monetary value and cost very little to acquire. And they have, for me, a visual appeal which I suspect was never intended by their anonymous makers.

In contrast to works of art they are wonderfully indifferent to my view of them. And because they do not strive to please, my pleasure in their

Engine rocker shaft
Controls the movement of valves in an engine.
29 cm/11½ in.

Cast-iron screw thread
A loose hook is attached to a bar in the hollow centre of this object but it is difficult to see how it was held in order to be turned.
17 cm/6¾ in.

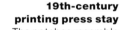

19th-century printing press stay
The notches resemble those on the bottom of a deck chair. Possibly from an etching press.
21 cm/8¼ in.

mysterious elegance is more intense than the pleasure I derive from art, if only because no artist, no conscious manipulator of my aesthetic response, stands between us. Although I did not create any of these objects, my discovery of them and recognition of their appeal allows me to experience a form of creative excitement that, though it may not match the excitement of an artist, is, in my experience, superior to the passive pleasure of a mere spectator.

In choosing to collect man-made things I do not deny the sculptural quality of natural objects. For many years my favourite weekend walk was along the south coast of England from St Margaret's Bay to Deal. To the exciting possibility of death by drowning or landslide was added the beachcomber's thrill of finding what looked like original works by Alberto Giacometti or Henry Moore protruding from the crumbling chalk cliff. Though bigger was better, it was also heavier, so difficult decisions had to be made that balanced weight, distance, time and aesthetics against the approaching tide. Flints that looked so pristine on the seashore now gather mould on my London garden walls. Some made it to the First Division of my mantelpiece and give pleasure still.

The objects illustrated here were found on a different kind of beach: the huge fairs, actually of junk but glorified by the name of 'antiques', held on deserted airfields or agricultural showgrounds upon which, every two months, hundreds of white vans converge. The vans disgorge the jumbled stock of dealers who have persuaded the heirs of the recently deceased to pay them to clear their sheds and attics. Spread out on the gravel and the grass is the leisure equipment of a pre-television age: croquet mallets, foils, golf clubs, bows and arrows, polo sticks, wind-up gramophones, sewing sets, woodworking tools, sheet music, fish tanks … a limitless sea of dross in which it is my task to find a few extraordinary things.

As I walk between the stalls my hunter's eye scans thousands of objects every minute, looking for that arresting combination of interesting shape and puzzling function, of elegance and mystery, that is instantly identifiable. It is like finding a friend's face in a crowd of strangers. The pleasurable shock of recognition mixes with the adrenal rush that accompanies any form of physical exercise. The eye, too, is a muscle, and for collectors the exercise of eye and brain involved in the search for their quarry is as stimulating and addictive as a gym workout or a game of tennis.

The collector's eye is not unlike an artist's. One needs to be alert yet passive, to use the peripheral

Adjustable mole clamp
The 'grip' of the jaws can be adjusted by turning the knurled wheel at the top of the handle.
19 cm/7½ in.

Wooden pattern
For casting machine parts
50 cm/19¾ in.

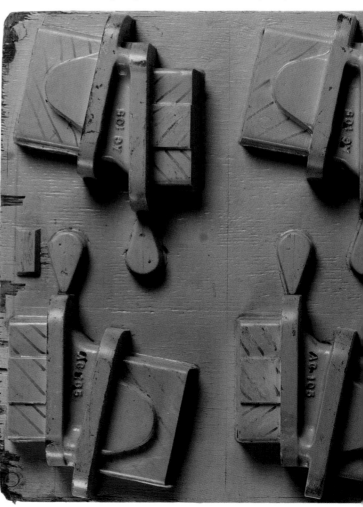

and instinctive rather than the rational and intellectual parts of one's brain. Yet there is a difference. Artists who incorporate found objects in their work seem to me to be engaged in a form of theft. They are kidnapping abandoned objects in order to give them a new identity as works of art. The co-opted scraps in a work by Joseph Cornell or Peter Blake are not of much interest in themselves. What is significant is the use to which the artist has put them and the fresh signature on the birth certificate. Marcel Duchamp's *Fountain* is, at one level, merely a mundane piece of sanitary ware, revered only as a consequence of the addition of a signature.

By contrast, the objects in this collection are attractive to me precisely because of their modest anonymity. No artist's signatures or designer labels here. No ambitious scholars or curators can

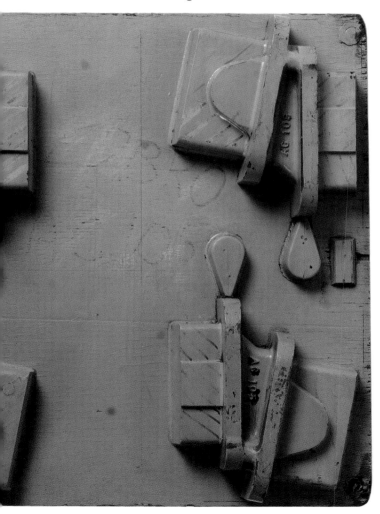

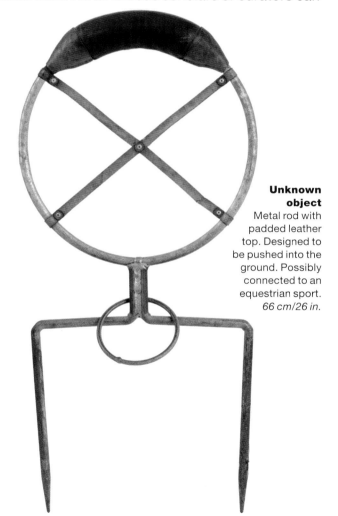

Unknown object
Metal rod with padded leather top. Designed to be pushed into the ground. Possibly connected to an equestrian sport.
66 cm/26 in.

hope to make a name by studying or exhibiting them. They do not need to be insured, protected from sunlight or stored in climate-controlled vaults. Burglars ignore them and they are too humble to be faked. Thankfully, they are of little interest to the army of dealers, auctioneers and art advisors who service and feed off wealthy collectors, helping them to buy and sell art as though it was no different from any other commodity.

Though I am critical of some aspects of the world of art, I do acknowledge a debt. Without the work of Moore, Giacometti, Brancusi, Jean Arp, Anish Kapoor and, in particular, the Surrealists, who opened my eyes to the mystery and beauty of the 'found object', it is doubtful that I would have noticed, let alone wanted to own, any of these things. Paradoxically, what attracts me to them is what sets them apart from genuine works of art – their indifference to my view of them. All works of art seek to engage in some way, to charm, please and entrance or to confront, challenge and enrage the viewer. These things couldn't care less. They were made to be used, not to be admired for their aesthetic qualities, and we are free to enjoy them in any way we choose.

The early Modernists who collected primitive masks and figures enjoyed a similar freedom. They could pillage them for formal devices without feeling any particular need to study or understand the societies from which they came. But the art of tribal societies is still art; made to be looked at, albeit in a different way from that in which Western collectors saw it.

The objects collected here have, thankfully, no aesthetic or social contexts that we need to understand in order to enjoy them. We can read what we like into them without fear of treading on tribal toes or offending the sensibilities of another culture. I can look at the surgical instruments on page 54 and, though shuddering at the use to which they are put, still enjoy the suggestion of dancers pirouetting to music.

This is the final and most important criterion for the selection of these objects: that each is suggestive of some natural form. In order to enjoy them fully the viewer needs to search their memory for images of faces, figures, animals or plants for which they stand as metaphors.

Some years ago there was a vogue for picture-books in which the illustrations seemed to consist of nothing but a colourful soup of tiny pixels. However, by deliberately crossing one's eyes and resetting their focus, the random pattern resolved itself into images of astonishing depth.

To enjoy this collection the viewer is advised to reset their brain to metaphorical mode. Like the *images-devinettes*, puzzle pictures popular at the end of the 19th century (is it a rabbit or a duck, a young girl or an old crone?), these are three-dimensional puzzles. Doubly so, in that once we have solved the problem of their function we can search them for hidden meanings and allusions, magical keys to the locked rooms of our unconscious. If all you see on pages 110 and 106 are a ceramic food mixer and the metal blade from a kitchen waste-disposer, you are missing the point. If, however, you are reminded by the first of Isadora Duncan in full flight or a cone of soft ice cream, and by the second of Harold Edgerton's multiple-exposure photographs of golfers or Leonardo's designs for a perpetual motion machine, you are definitely using your metaphorical eye.

The exercise of 'seeing as…' is all the more pleasurable because it is not prompted or controlled by an intermediary. No artist is telling you how to see. You are using your own eyes and a brain stocked with your own set of images and memories. You are walking unaided, without the Zimmer frame of art.

It is perhaps worth warning the sensitive reader that a proportion of the visual and verbal allusions are sexual. If you have a vivid sexual imagination, and the Surrealists were happy to admit that they did, there is material here for a museum of erotic metaphors. This has to be more entertaining than the boringly literal and descriptive Sex Museums of Paris, Amsterdam, Hamburg or Barcelona. When this book was no more than a collection of images, potential publishers argued that there needed to be a definite structure to help readers to make sense of what they saw. Could the objects be organized in some logical way: by trade, by material, by shape, date or size?

This is a problem common to all curators, whether they look after a national museum or a collection of stamps. It was painful at first to try to impose order on objects that had been plucked from the joyful disorder of the jumble stall and which are displayed in my flat with a deliberate lack of logic. Several systems of organization

Dog muzzle
Leather straps with adjustable collar.
15 cm/5¾ in.

Dentist's model jaws
Opposite: Dentists take an impression of the patient's upper and lower jaws.
13 cm/5 in.

were tried but, inspired by the Pitt Rivers Museum, it was sorting by function that produced both a manageable number of chapters and a stimulating mix of shapes within each category.

The conclusion for design theorists is the rather obvious one that though, for example, hitting and cutting are processes common to many trades, what determines the shape and the material of the tool used is the substance or material to which it is applied. To take an example: needles are used to make and repair nets, thatch, saddles, shoes, carpets and, in emergency, human flesh. They can be made of wood, bone, metal or plastic. Is it any wonder that they come in a variety of shapes and sizes?

By necessity, the categories of function chosen have been endowed with a certain elasticity. The chapter entitled 'Spreading', which groups fly-whisks and syringes with radiators and surgical dilators, is the one most likely to raise eyebrows.

Aluminium tongs
Right: For gripping tomatoes prior to slicing.
18 cm/7 in.

Electric coffee percolator
Far right: The legend on the Bakelite handle reads: 'Appareillage electrique moderne "MAGIC EXPRESS" Paris Brevete'. The top section lifts out a cylinder for water in which the coffee holder sits. The bottom two thirds of the percolator hold the heating element.
25 cm/9¾ in.

In the end, it was surprising that so much of a collection that was assembled without thought for these categories managed to fit them.

The quality of mystery that originally attracted me to these tools has, in most cases, been dispelled. I think I know how most of them were used. Some captions, however, remain speculative. Photographs of ten of the most puzzling were sent to the curators of the Science Museum, who could not identify any of them. So, if I have made any mistakes, at least I am in distinguished company. Corrections and suggestions are welcome.

Because the objects were photographed separately and then reduced and enlarged to fit their pages, the scale of adjacent images often differs. To give the reader an idea of scale, each caption records the major dimension of each object or, in the case of a group of similarly sized objects, a rough average.

Aluminium tongs
Left: For picking up cooked meat or fish.
16 cm/6¼ in.

Small coffee percolator
Far left: For making a single cup of coffee.
16 cm/6¼ in.

Ice tongs
The most efficient way to carry blocks of ice.
28 cm/11 in.

'To see is an act;
the eye sees as the
hand grasps.'
Paul Nougé Surrealist

'Beware of domestic objects.'
Claude Cahun, photographer

'To me, the tools used to make a
Fabergé egg are more interesting, perhaps
even more beautiful, than the egg itself.'
Robert Hughes, critic

'I like films which allow me to dream, not
those which impose their dreams on me.'
Georges Franju, film director

'The world is full of sources
but it is the artist who makes
the work of art.'
David Smith, sculptor

Hitting
To strike, tap, beat, batter, punch, stab, slap, bash, knock, swat or smash
Page 22

Cutting
To slice, chip, hack, drill, saw, lop, prune, trim, mill, split, mow, chop, slit or slash
Page 40

Gripping
To clutch, squeeze, grasp, pinch, nip, grab, clasp, catch, clamp or clench
Page 66

Holding
To carry, prop, trap, catch, contain, cup, cradle, enclose or support
Page 86

Rubbing
To roll, mix, comb, crush, grind, mash, scrape, massage, mince or whisk
Page 106

Shielding
To guard, protect, cover, screen, hide, deflect, insulate, mask or defend
Page 122

Moulding
To cast, fashion, bend, form, press, forge, shape, stamp or stretch
Page 132

Spreading
To expand, diffuse, disperse, circulate, dilate, open, scatter or transmit
Page 150

Testing
To check, calibrate, survey, mark out, estimate, gauge, judge or measure
Page 170

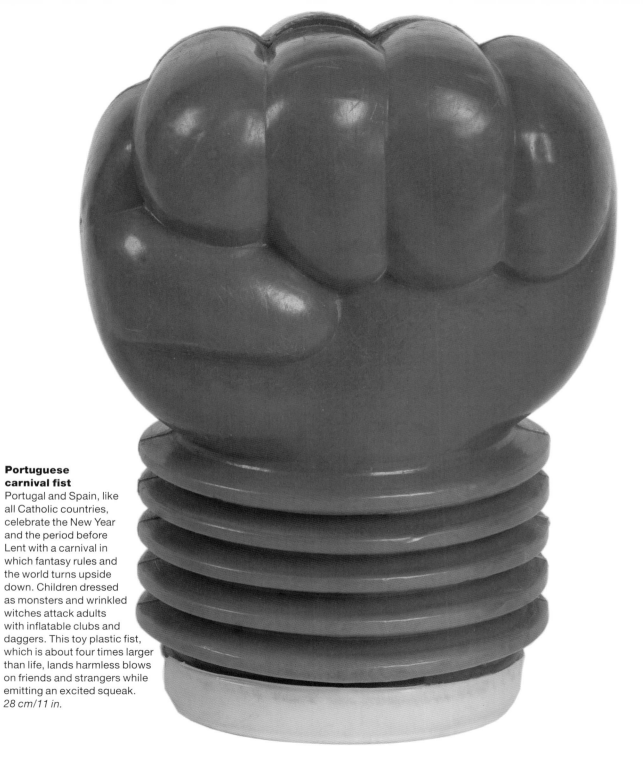

Portuguese carnival fist
Portugal and Spain, like all Catholic countries, celebrate the New Year and the period before Lent with a carnival in which fantasy rules and the world turns upside down. Children dressed as monsters and wrinkled witches attack adults with inflatable clubs and daggers. This toy plastic fist, which is about four times larger than life, lands harmless blows on friends and strangers while emitting an excited squeak.
28 cm/11 in.

Hitting
To strike, tap, beat, batter, punch, stab, slap, bash, knock, swat or smash

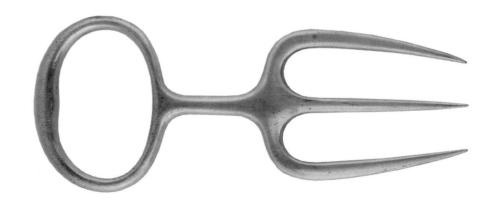

Carving fork
For holding hot roasts while they are being carved.
12 cm/4¾ in.

Anthropologists agree that man's first tools were hitting tools. Even apes and monkeys, with whom we share distant ancestors, have mastered the art of cracking nuts with rocks.

The common denominator of the tools in this chapter is the single, percussive blow. They are used to strike, among other things, flesh, fish, ice, soil, plants, carpets and thatch. The shape of each tool reflects both the material of which it is made and the particular task that it performs.

As in the subsequent chapters, the reader must accept that the common theme is a somewhat loose fit for a group of objects that were selected for their interesting shapes rather than as a comprehensive catalogue of all tools sharing this function. The 'hitting' theme has, for example, been expanded to include a group of passive tools, portable anvils used by farmers and shoemakers, whose function is to return the force and thus increase the efficiency of the hammer's blows. Where hitting

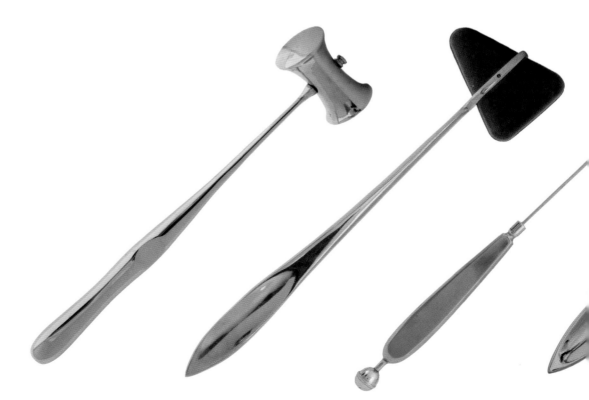

tools such as meat cleavers or eel spears have been given sharp edges, it could be argued that they have become cutting tools. In general, however, the single-blow definition holds.

More than half of these are hand tools, and it is interesting to note the way handle shapes are adapted to the task. The gardener's dibber and the cook's meat tenderizer are shaped for pulling as well as comfortable pushing. Clubs and other tools, like the butcher's cleaver or the fishmonger's ice pick, that are swung before striking, often have flared handles to help the user keep a firm grip.

Readers with a taste for visual puns are challenged to find a maternal gorilla, an Easter Island statue, a dancer *en point* and Queen Boadicea's chariot. Sculptors whose work helped me to see these objects or who might have been inspired by similar things include Constantin Brancusi, William Turnbull, Claes Oldenberg and Alberto Giacometti.

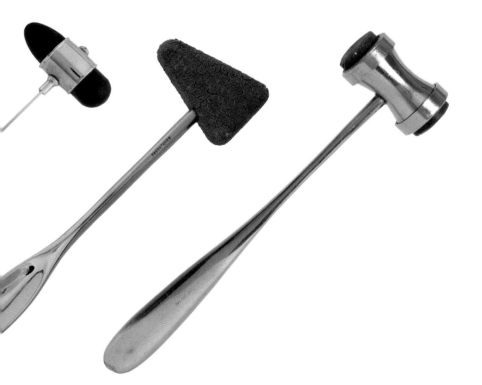

Five medical hammers
For testing muscular reflexes. Some of these have pointed handles for checking the nerves in damaged limbs.
All approx. 18 cm/7 in.

Hitting

From left to right:
African axe blade
The wooden handle was positioned midway between the pointed and the flared tips.
69 cm/27¼ in.

'Panga' knife from the Philippines
Found with a set of 19th-century English gardening tools. For hacking through undergrowth.
65 cm/25½ in.

Combination hammer and screwdriver
The hammer head makes a good handle for use as a screwdriver, but the shaft does not give much grip when used as a hammer.
33 cm/13 in.

Millwright's pick blade
Used to cut the grooves in millstones for milling flour.
18 cm/7 in.

Caulking iron
Used to press oakum or hemp into the gaps between the planks of wooden boats to make them watertight.
17 cm/6¾ in.

Fishmonger's ice pick
Used to break up blocks of ice and to manoeuvre large fish on the slab. The swollen foot of the handle prevents the shaft from slipping through the fishmonger's cold, wet fingers.
29 cm/11½ in.

Hitting

Six tribal clubs from islands in the Pacific
The simple objects produced by tribal societies, such as clubs, headrests, cooking utensils and musical instruments, often have a greater sculptural appeal than more obviously decorative items such as masks and carved figures. From left to right, these come from: the Solomon Islands, Samoa, Tonga, Hokkaido, the Gilbert Islands and Fiji. The first and third have been designed with dangerous points at both ends. The club from Hokkaido, a Japanese island, is identical to the clubs wielded by the Farnese Hercules or the Cerne Abbas Giant. The Gilbert Islanders, having no hardwoods, gave their clubs and spears a cutting edge by sewing onto them rows of tiny shark's teeth.
All approx. 96 cm/37¾ in

Hitting

Cast-iron cobbler's anvil
Most anvils for making or repairing shoes are made either to stand on a surface or to slot into a solid wood base, thus leaving both the cobbler's hands free. This anvil, found in a thrift shop in Tennesse, USA, could only be used if held inside the shoe with one hand, while the other wielded the hammer. Most cobbler's hammer heads were magnetized so that the cobbler could, using only one hand, hammer in nails that he kept in his mouth.
22 cm/8¾ in.

Small double anvil
Two large nails welded together. When one is being used the other prevents the anvil from disappearing into the ground.
22 cm/8¾ in.

Hitting

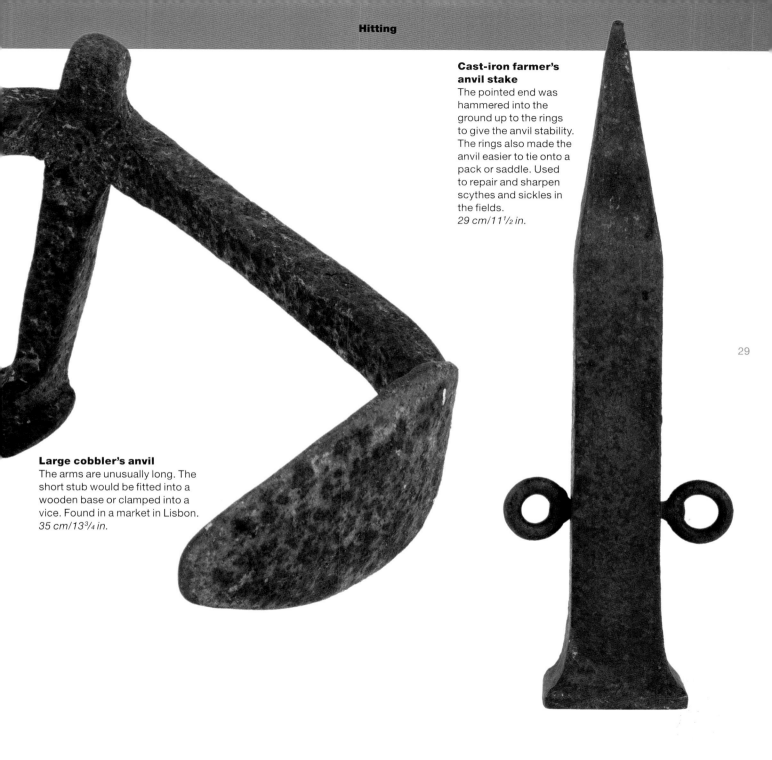

Cast-iron farmer's anvil stake
The pointed end was hammered into the ground up to the rings to give the anvil stability. The rings also made the anvil easier to tie onto a pack or saddle. Used to repair and sharpen scythes and sickles in the fields.
29 cm/11½ in.

Large cobbler's anvil
The arms are unusually long. The short stub would be fitted into a wooden base or clamped into a vice. Found in a market in Lisbon.
35 cm/13¾ in.

Hitting

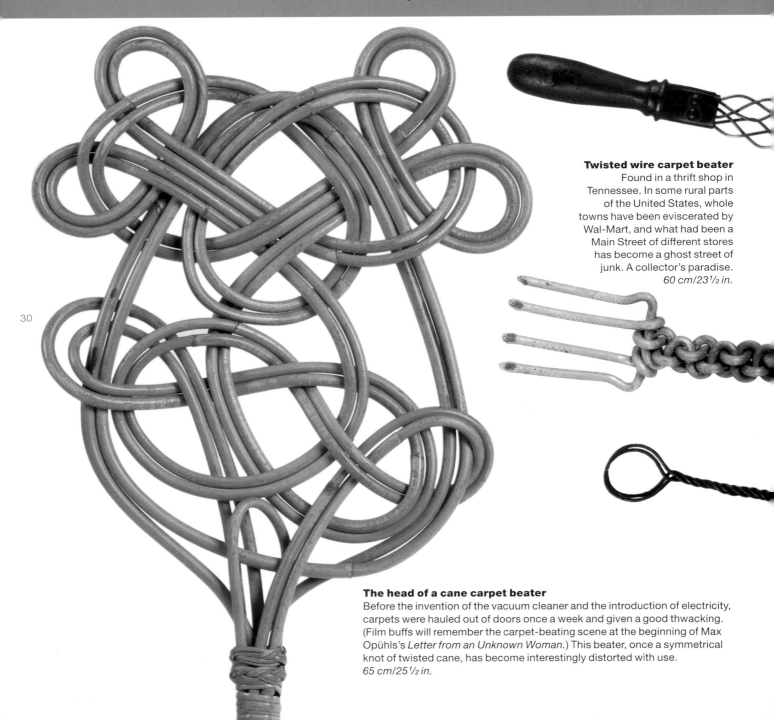

Twisted wire carpet beater
Found in a thrift shop in Tennessee. In some rural parts of the United States, whole towns have been eviscerated by Wal-Mart, and what had been a Main Street of different stores has become a ghost street of junk. A collector's paradise.
60 cm/23 1/2 in.

The head of a cane carpet beater
Before the invention of the vacuum cleaner and the introduction of electricity, carpets were hauled out of doors once a week and given a good thwacking. (Film buffs will remember the carpet-beating scene at the beginning of Max Opühls's *Letter from an Unknown Woman*.) This beater, once a symmetrical knot of twisted cane, has become interestingly distorted with use.
65 cm/25 1/2 in.

Hitting

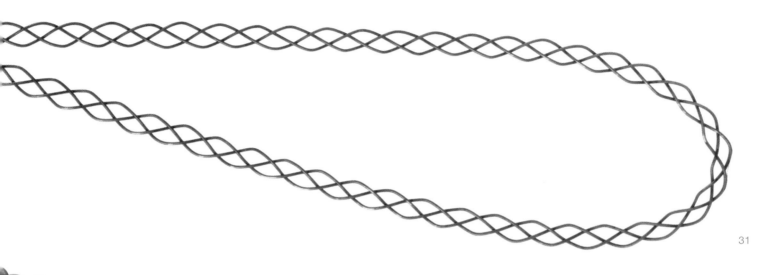

Two twisted wire toasting forks
Braiding or twisting the shaft of a toasting fork allows the heat to dissipate. The bottom fork is made of steel wire and the top fork is made of aluminium wire. These were products, like the kitchen implements on pages 98–9, of the tinkering industry.
Both approx. 32 cm/12½ in.

Two stone 'celts'
Elegant stone tools of unknown origin, possibly from New Guinea. Shaped by abrasion rather than knapping, these ancestors of the Stanley knife could have been held in the hand and used for cutting and scraping, or lashed to a wooden shaft and used as axe heads.
Top: 24 cm/9½ in.
Bottom: 32 cm/12½ in.

Hitting

Butcher's cleaver
A hybrid of a hammer and an axe, cleavers do the heavy work of dismembering a carcass. Designed to be used on a butcher's block, both ends of the blade are used for smashing and cutting through bone and sinew. The handle is flared to give a firm grip.
31 cm/12 1/4 in.

Hitting

Three seed planters
Bottom right: Known as a 'dibber' in England, this hand tool is used by gardeners to punch conical holes in the soil.
28 cm/11 in.
Right: This tool, which has lost its wooden handle, is long enough to be used without bending and heavy enough to make a hole just by being dropped.
67 cm/26¼ in.
Left: This American example is rather more ingenious in design. The hole is made by stabbing the soil with the metal-covered tips of two parallel arms, which, as they are withdrawn, act as a funnel for the seed released from the small gap near the handle.
110 cm/43¼ in.

Head of a garden rotivator
This has two propeller-like blades angled towards each other, which, when attached to a wooden shaft, would be pushed over the ground to open up the soil. Note the similarity of blade shape to the waste-disposal blade on page 106.
23 cm/9 in.

Garden rake or carding tool
Judging by the rust and corrosion, this must be at least 200 years old. Hand-forged and curiously delicate for a garden tool, the tines resemble the spines of the hackle board on page 121, suggesting that this may have been the head of a tool for combing flax prior to spinning.
18 cm/7 in.

Hitting

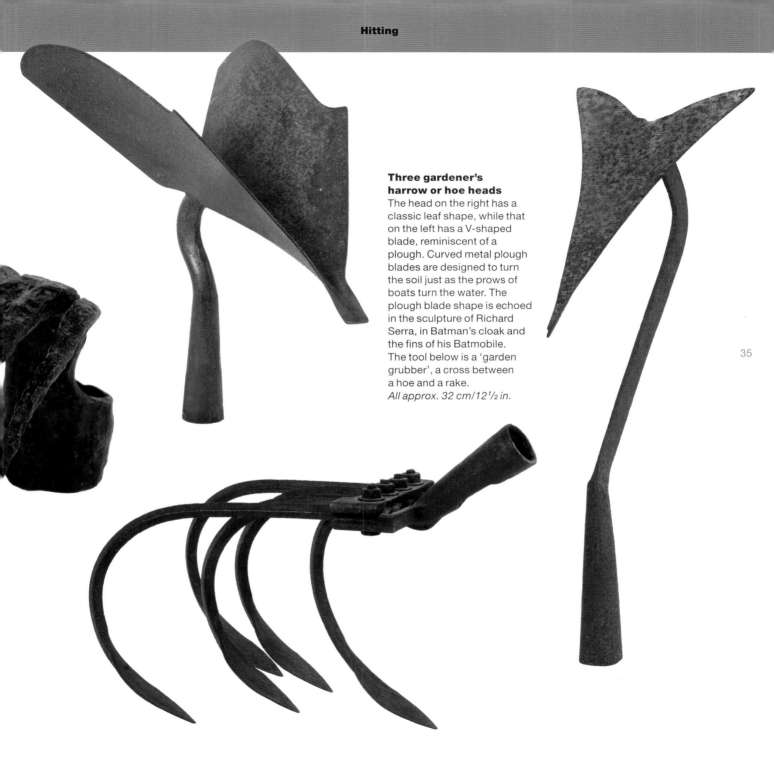

Three gardener's harrow or hoe heads
The head on the right has a classic leaf shape, while that on the left has a V-shaped blade, reminiscent of a plough. Curved metal plough blades are designed to turn the soil just as the prows of boats turn the water. The plough blade shape is echoed in the sculpture of Richard Serra, in Batman's cloak and the fins of his Batmobile. The tool below is a 'garden grubber', a cross between a hoe and a rake.
All approx. 32 cm/12½ in.

Hitting

Farmer's needle
The threaded tip of this elegant metal needle can be pushed through a metre-thick bundle of straw to tie it into a bale.
109 cm/43 in.

Metal net needle
The string for repairing nets was wound around and between the central spike and the two legs and unwound as needed. Net needles are usually made of wood.
26 cm/10¼ in.

Thatcher's needle
With the needle retracted, the hook is thrown over a bundle of reeds (the wooden handle is missing). The needle is then pushed through the reeds and pulled back carrying the string with which the bundle is tied.
41 cm/16 in.

Hitting

Two eel spears
This page and opposite page, centre: Usually made of wrought iron, eel spears, or 'gleaves', are made in a variety of shapes. The functional problem is that of securing the eel without cutting it in half. The example opposite, from 19th-century France, is shaped like a classical trident. The mask-like example on the right is English and of a much later date.
Opposite: 26 cm/10¼ in.
Right: 60 cm/23¼ in.

French tanner's hook
For lifting and dragging wet hides. The shape and covering of the handle help the tanner to grip the hook.
20 cm/7¾ in.

A double bale hook
Used to grip and manhandle bales of hay and cotton or large wooden barrels. Traditional docker's hooks used for working and fighting had one hook and a wooden handle.
23 cm/9 in.

Hitting

Garden rotivator
Attached to a long handle, this perforates and lifts the soil prior to sowing.
23 cm/9 in.

Hitting

Meat tenderizer
The spiked tips can be unscrewed and reversed into the plastic cup for safe storage.
14 cm/5 1/2 in.

Wooden model aeroplane propeller
Propellers were iconic objects of Modernism. The wooden models of prototype ship's propellers in London's Science Museum rival anything in the nearby Victoria and Albert Museum. One of the first objects acquired for the collection of the Department of Industrial Design at the Museum of Modern Art in New York was a ship's propeller.
29 cm/11 in.

Cutting
To slice, chip, hack, drill, saw, lop, prune, trim, mill, split, mow, chop, slit or slash

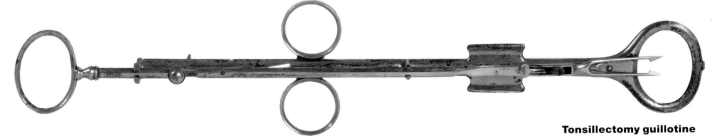

Tonsillectomy guillotine
A slightly different design to the guillotine on page 50. The finger holes resemble those on the syringe on page 154.
18 cm/7 in.

Cutting

Cutting tools logically develop from hitting tools. Prehistoric hunters who tried to smash animal bones between two rocks may have found, when one of the stones split, that they had created a sharp-edged tool which could then be used to cut flesh. It was a short step from the use of accidentally broken stones to deliberately knapped and shaped knives, axes and arrowheads.

In New Mexico where I grew up in the 1940s the local Indians had only just stopped using stone tools, and any sharp-eyed boy could begin a lifetime of scavenging with a collection of flint arrowheads.

The tools in this chapter cut herbs, air, plants, flesh, bone, wood, grass, bread, pastry, nails, metal, plaster and cheese. Their shapes differ because the materials differ and because the cutting action changes to suit the material. Lawnmower blades and milling tools rotate, herb cutters rock, pasta cutters roll, castrators and nail scissors nip, saws rub, the axe and the guillotine fall.

Cutting and gripping tools are frequently expressive because the ratio of handle to jaws, giving the user mechanical advantage, results in many examples of striding legs and gaping mouths.

Cutting

Tree lopper
Attached to a wooden pole, this is operated by a rope running over a rotating sheave located at the end of the long arm. When the arm is pulled down, against the pressure of the leaf spring, a crescent-shaped blade swivels upwards to cut the branch held by the sharpened hook.
27 cm/10½ in.

Cutting

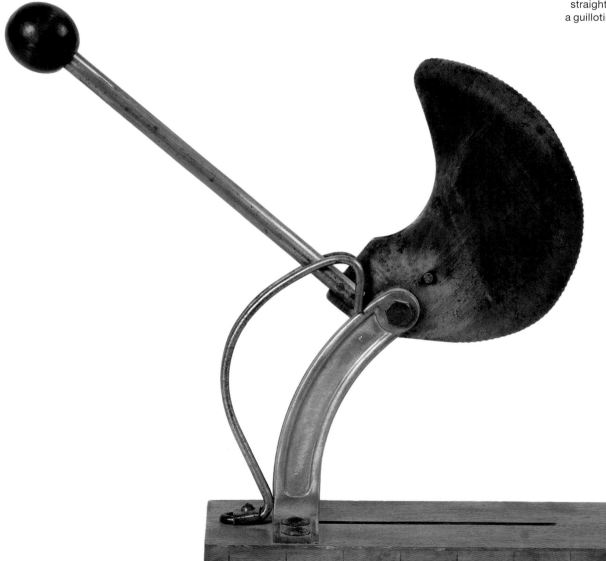

Baguette cutter
The curved blade slides through the bread, cutting it more effectively than a straight edge. The blade of a guillotine was diagonal for the same reason.
34 cm/13¼ in.

Cutting

Adjustable pasta cutter
The rolling blade is also used in the traditional pizza cutter.
8 cm to 40 cm/3 in. to 15³⁄₄ in.

Cutting

A pair of French vine secateurs
The wooden handles are missing. The blades are curved, like those on the bone shears on page 50, to ensure a better grip on whatever is being cut.
49 cm/19¼ in.

Animal castrator
A standard item of veterinary equipment. The blunt blades compress but do not cut. The double pivot mechanism is also employed on bolt cutters and on the folding barbed-wire cutters used by the military.
32 cm/12½ in.

Cutting

Hoof trimmer
Found in the yard of an abandoned farm. The blades can be unscrewed for sharpening.
30 cm/11¾ in.

Cutting

A swan's neck mortise chisel
Used to clean out rectangular holes in doors so they can receive mortise locks, and also to clean the rectangular spoke holes in the hubs of cartwheels. The curve in the neck acts as a pivot to allow the tip of the blade to cut upwards in a confined space.
35 cm/13¾ in.

Carpenter's auger
The wooden handle, to which this was attached, has been burnt off.
44 cm/17¼ in.

Cooper's auger
For cutting conical bungholes in barrels. The missing handle was a wooden dowel threaded through the sleeve at the top, which gave the cooper maximum torque.
28 cm/11 in.

Florist's secateurs
The central spring keeps the tool open and gives the user better control of the cutting action. The function of the little blade at the end of one of the arms is unknown.
15 cm/5¾ in.

Cutting

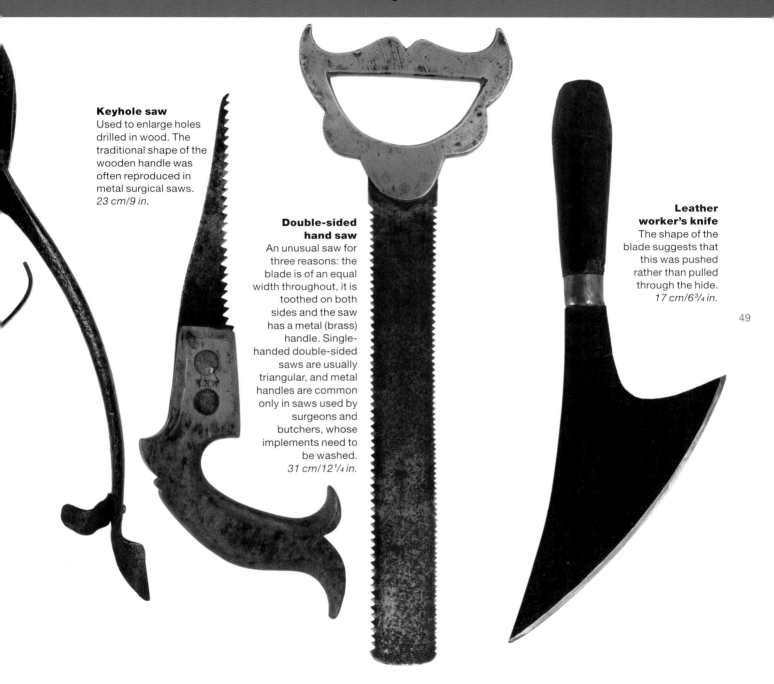

Keyhole saw
Used to enlarge holes drilled in wood. The traditional shape of the wooden handle was often reproduced in metal surgical saws.
23 cm/9 in.

Double-sided hand saw
An unusual saw for three reasons: the blade is of an equal width throughout, it is toothed on both sides and the saw has a metal (brass) handle. Single-handed double-sided saws are usually triangular, and metal handles are common only in saws used by surgeons and butchers, whose implements need to be washed.
31 cm/12¼ in.

Leather worker's knife
The shape of the blade suggests that this was pushed rather than pulled through the hide.
17 cm/6¾ in.

Cutting

Tosillectomy guillotine
As the shaft is pushed forward the two-pronged lance spears and lifts the tonsil so that it can be cut by the blade that follows.
18 cm/7 in.

Bone shears
The two arms can be slid apart for disinfection.
15 cm/5¾ in.

Plaster shears
Described in medical instrument catalogues as 'Lorenz's plaster shears', these, like the long-handled shears on page 54, needed to be carefully designed so that they cut through the plaster cast but not the patient's flesh. Did surgeons earn royalties from the sale of instruments that they had designed?
23 cm/9 in.

Cutting

Food masticator
A pair of scissors used to cut up food for patients with no teeth. The arms close against the force of a coiled leaf spring.
17 cm/6¾ in.

Turbine blades
Two blades from a small turbine. These are fixed to a hub using a serrated key similar to the one that holds the teeth of the mill blade on page 61.
15 cm/5¾ in.

Cutting

Plastic blade for a lawnmower
The same pattern that a whirlwind leaves after flattening a cornfield.
28 cm/11 in. diameter

Cutting

Plant secateurs
Try to cut a twig with a pair of scissors and the twig will slide out from between the blades. The mechanism of this tool pulls the upper blade downwards, thus ensuring that the plant stem cannot slide out of the jaws.
17 cm/6¾ in.

Shrub pruners
Fixed to a pole and operated by pulling a cord attached to one of the handles, this extends the reach of the gardener's arm.
29 cm/11½ in.

Cutting

Three surgical tools
Right: Described in catalogues as a *rongeur*, which is the French for rodent. 'Nibbler' would be an accurate English translation. The jaws are opposed cups with sharpened rims.
22 cm/8¾ in.
Centre: A perforator. With the handles open the two blades have a dagger-like profile with which the surgeon makes the first incision. Closing the handles causes the blades to open, like reverse scissors.
24 cm/9½ in.
Far right: Secateurs for cutting plaster casts. The parrot-beak jaws allow the blades to slide safely along the patient's skin.
34 cm/13¼ in.

Cutting

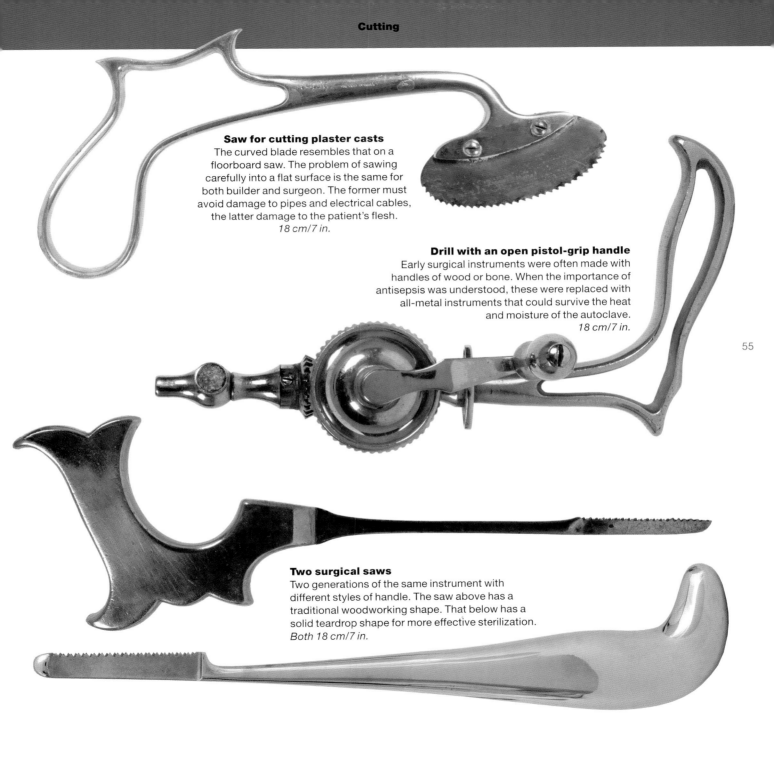

Saw for cutting plaster casts
The curved blade resembles that on a floorboard saw. The problem of sawing carefully into a flat surface is the same for both builder and surgeon. The former must avoid damage to pipes and electrical cables, the latter damage to the patient's flesh.
18 cm/7 in.

Drill with an open pistol-grip handle
Early surgical instruments were often made with handles of wood or bone. When the importance of antisepsis was understood, these were replaced with all-metal instruments that could survive the heat and moisture of the autoclave.
18 cm/7 in.

Two surgical saws
Two generations of the same instrument with different styles of handle. The saw above has a traditional woodworking shape. That below has a solid teardrop shape for more effective sterilization.
Both 18 cm/7 in.

Cutting

Two African knife blades
Both lack handles. The blade on the right is from a throwing knife and may have been used as currency.
Both 60 cm/26 in.

Axe head
Described by the vendor as an executioner's axe. More probably used to trim vines.
41 cm/16 in.

Cutting

Rusty hoe
At first sight this looks like a pizza oven paddle but research has established that it is the head of a Sudanese hoe.
21 cm/8¼ in. diameter

Cutting

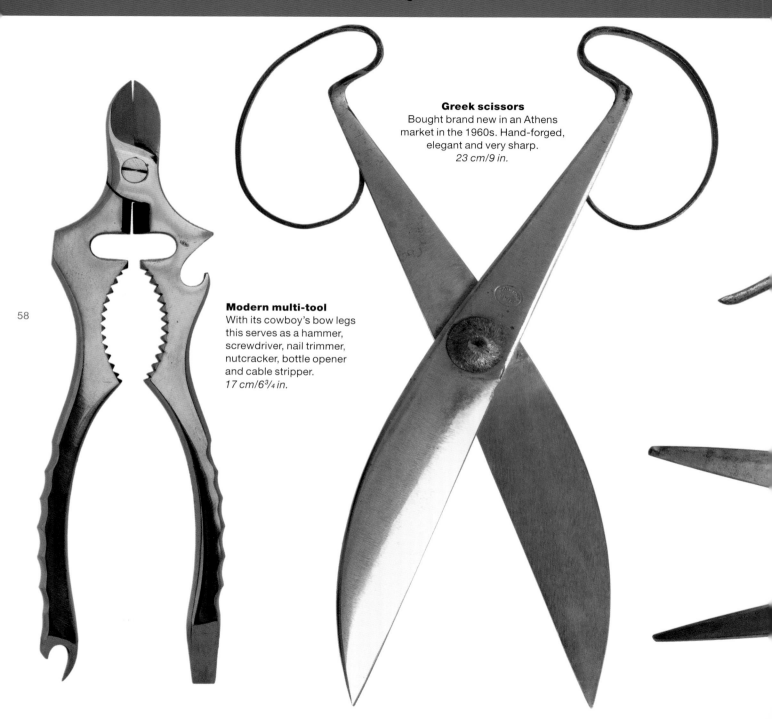

Modern multi-tool
With its cowboy's bow legs this serves as a hammer, screwdriver, nail trimmer, nutcracker, bottle opener and cable stripper.
17 cm/6¾ in.

Greek scissors
Bought brand new in an Athens market in the 1960s. Hand-forged, elegant and very sharp.
23 cm/9 in.

Cutting

Two pairs of tailor's scissors
Above: The loop of the handle on this more primitive pair has been opened to give the cutter greater power over the blades.
25 cm/9¾ in.
Below: Traditional tailor's scissors with ergonomically shaped, offset handles, which prevent this heavy tool from twisting in the cutter's hand as he slides it over the cutting table.
37 cm/14½ in.

Cutting

Three herb knives
Above: With four long blades, this is a tool for a very large kitchen. The two handles at right angles to the blades allow the user to apply greater pressure than the conventional vertical wooden handles.
30 cm/11¾ in.
Left: Two single-bladed herb knives. The function of the spike on the larger knife is unclear. All herb knives have curved blades to crush as they cut.
Far left: 17 cm/6¾ in. diameter.
Left: 21 cm/8¼ in.

Cutting

Steel mill blade
Lathes work by applying a sharp cutting tool to rotating metal. A mill passes the object to be shaped under a rotating cutter like a chunky circular saw. Worn teeth can be easily knocked out for sharpening.
15 cm/5¾ in. diameter

Cutting

Slater's rip
Roofing tiles are lapped over each other like fish scales. The nails attaching the tiles to the wooden battens beneath are therefore covered and inaccessible. To replace a broken tile or slate the roofer slides this tool under the damaged tile. When he has located the nail with the blunt mushroom-shaped tip, he moves the rip past it and then pulls sharply backwards, cutting the nail with the sharpened underside.
The offset handle resembles that of a cake knife, designed to slide under the cake and lift a slice.
59 cm/23¼ in.

Farrier's buttris
A tool for trimming horses' hoofs. The blade shape and offset shaft are similar to the slater's rip. This is essentially an adapted chisel. The wooden handle, resembling a gunstock, is pushed with the shoulder while the farrier's hand grips the offset and guides the blade.
39 cm/15¼ in.

French farmer's hay knife
The two metal spikes at right angles to each other originally had wooden handles, resembling the handles on a scythe. Hay knives are for cutting hay that has been compacted. The blades are usually serrated so that they can be used with a sawing action.
86 cm/33¾ in.

Cutting

Chip cutter
The grid of sharp blades can be placed over any hard-fleshed fruit or vegetable and pressed down to produce perfect square-sectioned chips.
26 cm/10¼ in.

Curd agitator
The head of a dairyman's tool, resembling a latticed shovel, used to cut and stir the mixture of milk and rennet that coagulates into junket – the first stage in the making of cheese.
Overall width 25 cm/9¾ in.

Rotary biscuit cutter
When rolled over flattened dough, the two counter-posed kidney-shaped blades cut perfect circles. This is quicker than pressing on the dough with an upturned glass.
17 cm/6¾ in.

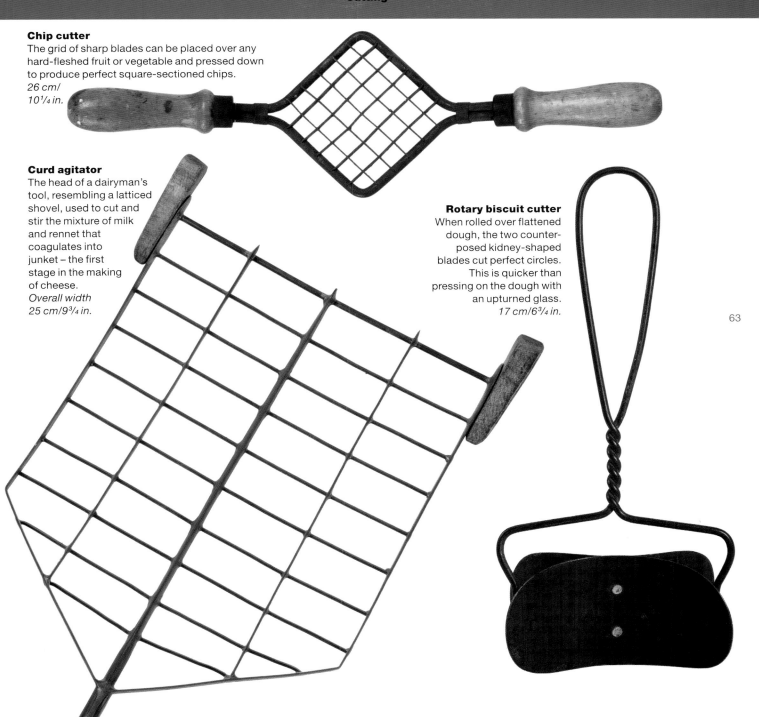

Cutting

Cork washer cutters
A set of nested cutters of different sizes. Each tube has a sharpened tip which is pressed into the cork sheet to cut out a disc. A smaller cutter is then used to make the central hole of the washer.
21 cm/8¼ in.

Mr and Mrs biscuit cutters
The difference in size and implied importance must have appealed to female cooks.
Left: 12 cm/4¾ in.
Right: 18 cm/7 in.

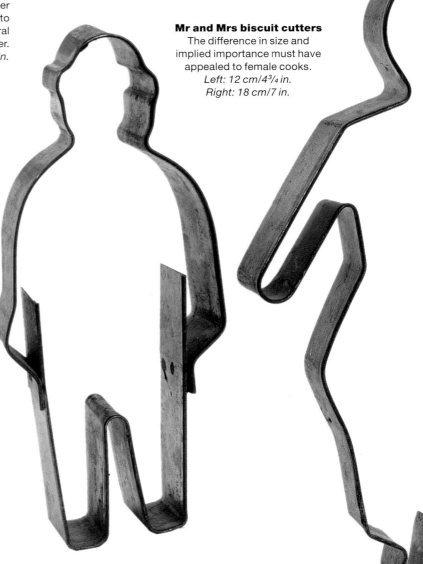

Cutting

Three devices for cutting eggs
Right: A tool for cutting the top off a boiled egg, for those too grand to use a knife or a teaspoon. Below left: A tool for slicing a hard-boiled egg into six equal segments. Below right: A similar tool, based on the principle of the guillotine.
All approx. 10 cm/4 in.

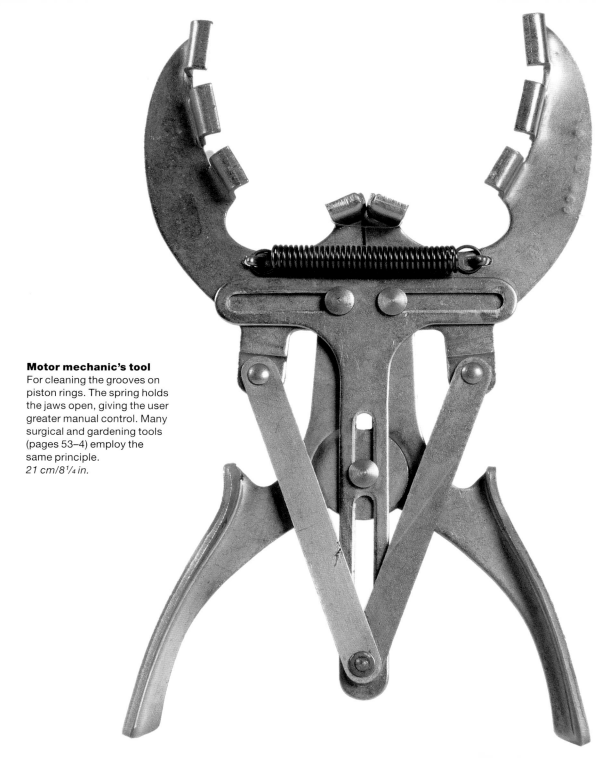

Motor mechanic's tool
For cleaning the grooves on piston rings. The spring holds the jaws open, giving the user greater manual control. Many surgical and gardening tools (pages 53–4) employ the same principle.
21 cm/8¼ in.

Gripping

To clutch, squeeze, grasp, pinch, nip, grab, clasp, catch, clamp or clench

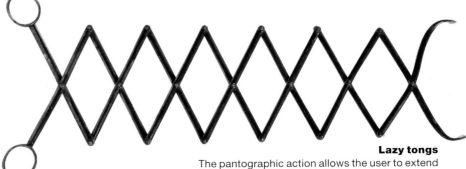

Lazy tongs
The pantographic action allows the user to extend this tool to pick up small, inaccessible objects or to hold food in front of an open fire.
6 cm to 60 cm/2¼ in. to 23½ in.

If hitting and cutting tools were the first steps in our ancestors' development of technology, gripping tools must have followed soon after. The efficiency of a hand holding a sharp stone was greatly increased if the stone blade was attached to a wooden shaft to make an axe or a spear, and the cutting-up of carcasses is much easier if you can grip the part you are working on.

Gripping is a natural function performed by most animals. Crocodiles' jaws, insects' mandibles, squids' suckers, parrots' beaks, lions' claws, bears' paws and monkeys' fingers are all essential for their survival. Gripping tools are a extension of our hands. You can thread a nut loosely onto a bolt with your fingers but you can only tighten it with a spanner. You can pick up a block of ice with your hands but you can't carry it far without it melting and slipping to the ground. Ice tongs are thermally and mechanically a better option. Try rearranging the logs on an open fire without the aid of fire tongs!

Gripping

The tools in this chapter help the user to manipulate things that are too small, cold, slippery, inaccessible or hot to handle. They grip skulls, nuts, bones, flesh, ice, vines and candles so that some action – cutting, twisting, pulling, lifting or snuffing – can be performed more easily.

It is no accident that this is one of the largest sub-groups in this collection, given the expressive nature of its function. The handles often suggest running legs, the gripping ends resemble shouting, grinning or glowering faces, while the closing of pivoting parts mimics the exchanges of couples locked in love, lust or battle.

As with previous chapters the thematic net is quite . elastic. The boundaries between gripping and holding are blurred. Does a snow shoe on page 75 grip the snow or hold its wearer on the soft surface? The medical inhaler on page 77 both holds a liquid and grips the invalid's face.

Hernia truss
The sprung steel core of this leather-covered truss grips and reinforces the stomach muscles of the wearer.
31 cm/12¼ in.

Gripping

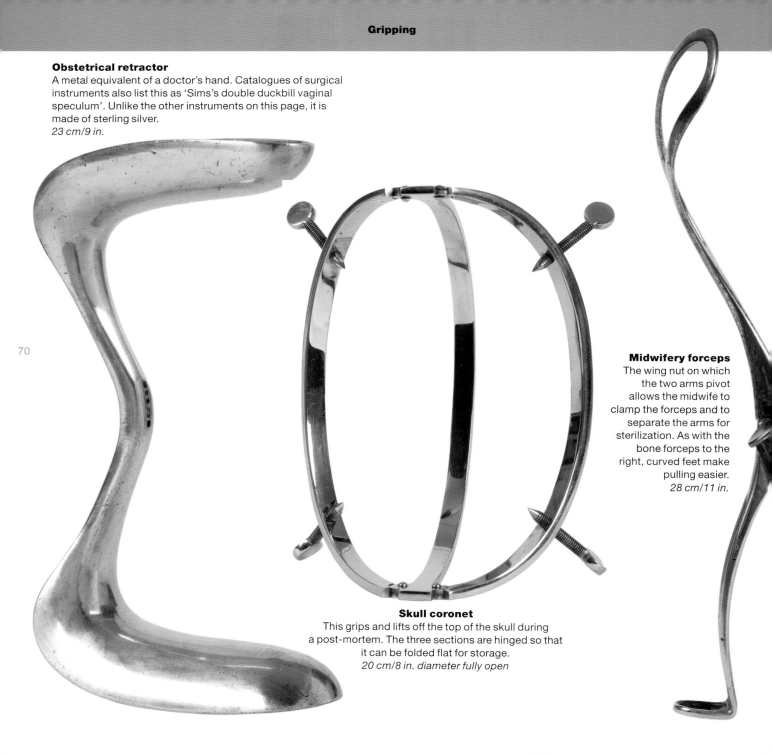

Obstetrical retractor
A metal equivalent of a doctor's hand. Catalogues of surgical instruments also list this as 'Sims's double duckbill vaginal speculum'. Unlike the other instruments on this page, it is made of sterling silver.
23 cm/9 in.

Midwifery forceps
The wing nut on which the two arms pivot allows the midwife to clamp the forceps and to separate the arms for sterilization. As with the bone forceps to the right, curved feet make pulling easier.
28 cm/11 in.

Skull coronet
This grips and lifts off the top of the skull during a post-mortem. The three sections are hinged so that it can be folded flat for storage.
20 cm/8 in. diameter fully open

Gripping

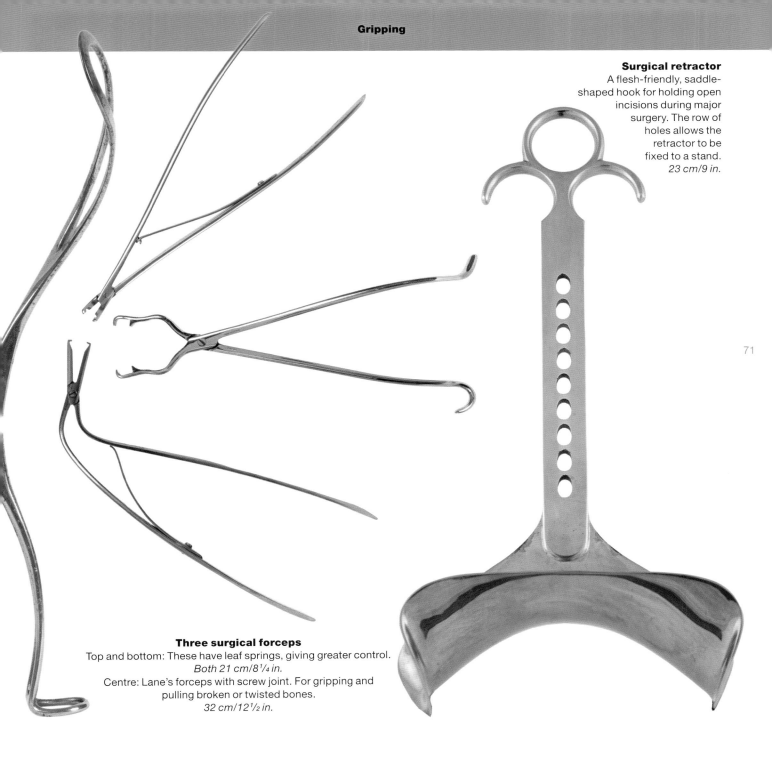

Surgical retractor
A flesh-friendly, saddle-shaped hook for holding open incisions during major surgery. The row of holes allows the retractor to be fixed to a stand.
23 cm/9 in.

Three surgical forceps
Top and bottom: These have leaf springs, giving greater control.
Both 21 cm/8¼ in.
Centre: Lane's forceps with screw joint. For gripping and pulling broken or twisted bones.
32 cm/12½ in.

Gripping

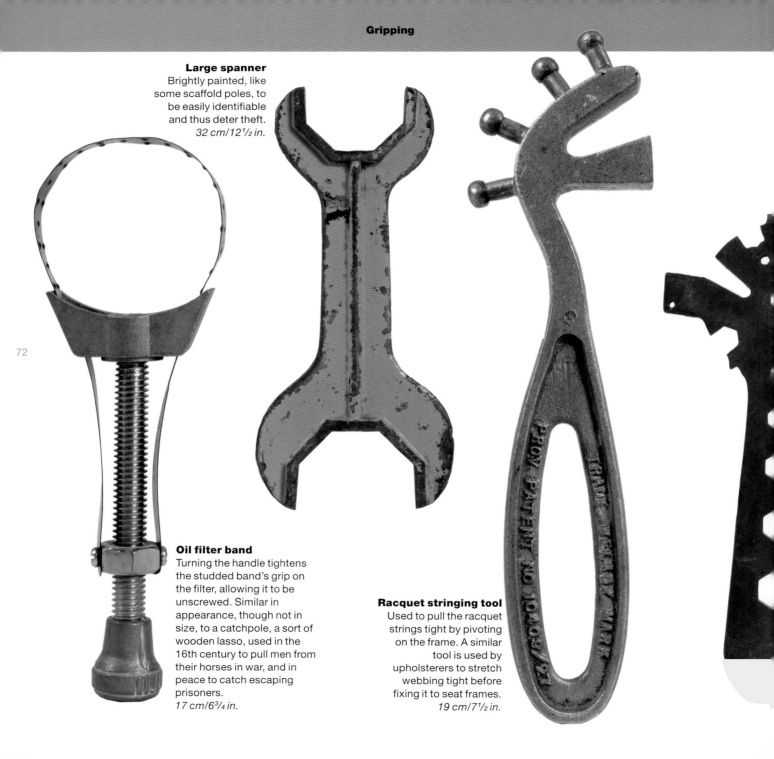

Large spanner
Brightly painted, like some scaffold poles, to be easily identifiable and thus deter theft.
32 cm/12½ in.

Oil filter band
Turning the handle tightens the studded band's grip on the filter, allowing it to be unscrewed. Similar in appearance, though not in size, to a catchpole, a sort of wooden lasso, used in the 16th century to pull men from their horses in war, and in peace to catch escaping prisoners.
17 cm/6¾ in.

Racquet stringing tool
Used to pull the racquet strings tight by pivoting on the frame. A similar tool is used by upholsterers to stretch webbing tight before fixing it to seat frames.
19 cm/7½ in.

Gripping

Motor mechanic's spanner
For tightening wheel nuts. Made by bending a tube with a hexagonal cross-section.
24 cm/9½ in.

Double-headed adjustable spanner
Turning the handle opens or closes the jaws.
27 cm/10½ in.

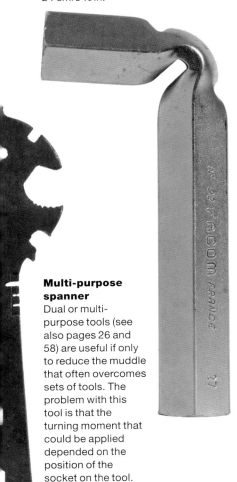

Multi-purpose spanner
Dual or multi-purpose tools (see also pages 26 and 58) are useful if only to reduce the muddle that often overcomes sets of tools. The problem with this tool is that the turning moment that could be applied depended on the position of the socket on the tool.
18 cm/7 in.

Dipping clamp
The jaws are held shut by a tension spring above the pivot, rather than a compression spring below it. Used to secure metal objects immersed in an electroplating tank.
18 cm/7 in.

Gripping

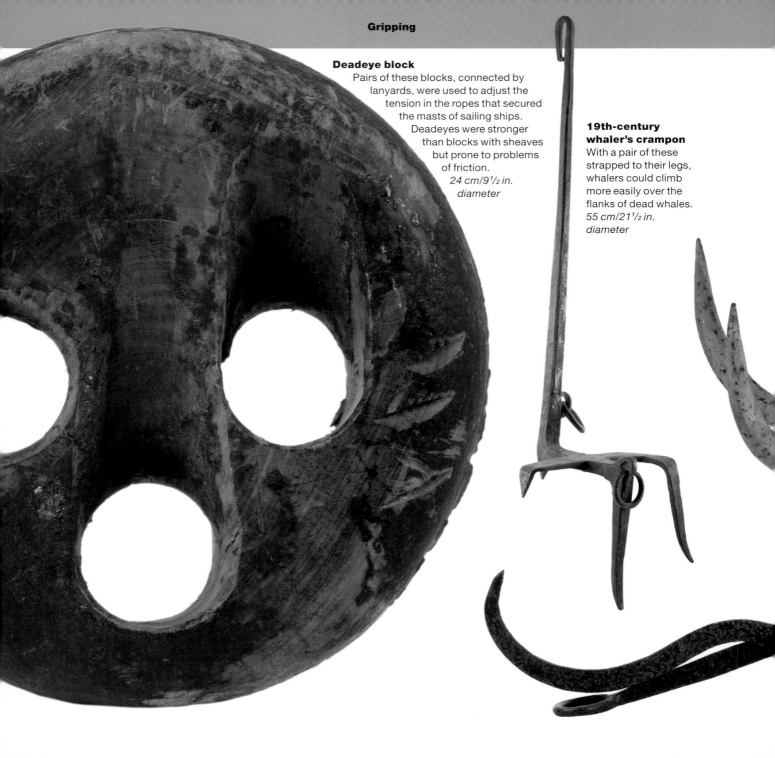

Deadeye block
Pairs of these blocks, connected by lanyards, were used to adjust the tension in the ropes that secured the masts of sailing ships. Deadeyes were stronger than blocks with sheaves but prone to problems of friction.
24 cm/9½ in. diameter

19th-century whaler's crampon
With a pair of these strapped to their legs, whalers could climb more easily over the flanks of dead whales.
55 cm/21½ in. diameter

Gripping

Portuguese anchor
Painted, like the spanner on page 72, to deter theft.
27 cm/10½ in.

Snow shoe
Snow shoes are usually flat with rounded ends. This has a pointed end, which curves upwards like the tip of a ski.
109 cm/43 in.

Ice tongs
A common tool before the advent of refrigeration.
80 cm/31½ in.

Ice axe
The hole in the blade allows the axe to be attached to the climber.
51 cm/20 in.

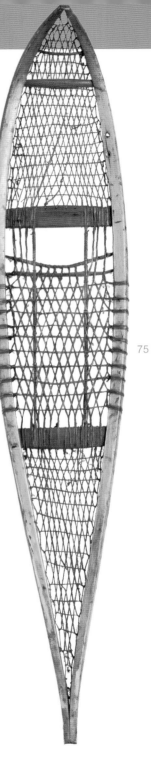

Gripping

Female urinal
Traditional item of hospital equipment. Ceramic items were later replaced by plastic, which has in turn been replaced by moulded paper equivalents.
26 cm/10¼ in.

Ceramic mug
Found at a Portuguese ceramics market. The hook gives a user with arthritis a better grip.
18 cm/7 in. diameter

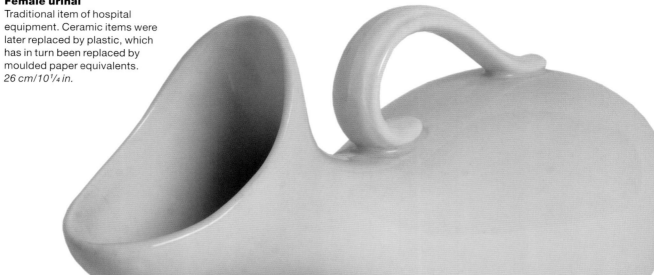

Gripping

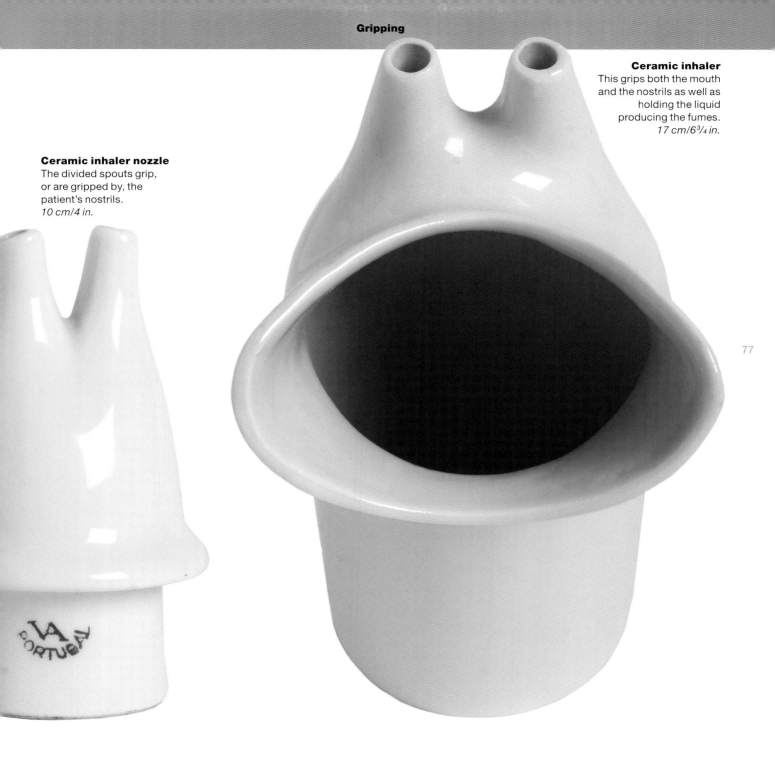

Ceramic inhaler nozzle
The divided spouts grip, or are gripped by, the patient's nostrils.
10 cm/4 in.

Ceramic inhaler
This grips both the mouth and the nostrils as well as holding the liquid producing the fumes.
17 cm/6¾ in.

Gripping

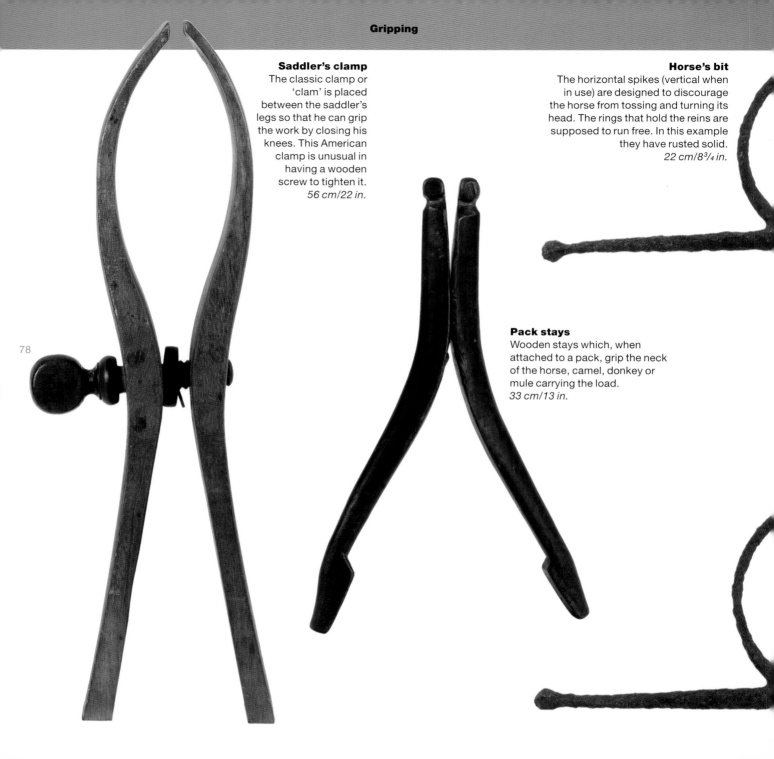

Saddler's clamp
The classic clamp or 'clam' is placed between the saddler's legs so that he can grip the work by closing his knees. This American clamp is unusual in having a wooden screw to tighten it.
56 cm/22 in.

Horse's bit
The horizontal spikes (vertical when in use) are designed to discourage the horse from tossing and turning its head. The rings that hold the reins are supposed to run free. In this example they have rusted solid.
22 cm/8¾ in.

Pack stays
Wooden stays which, when attached to a pack, grip the neck of the horse, camel, donkey or mule carrying the load.
33 cm/13 in.

Gripping

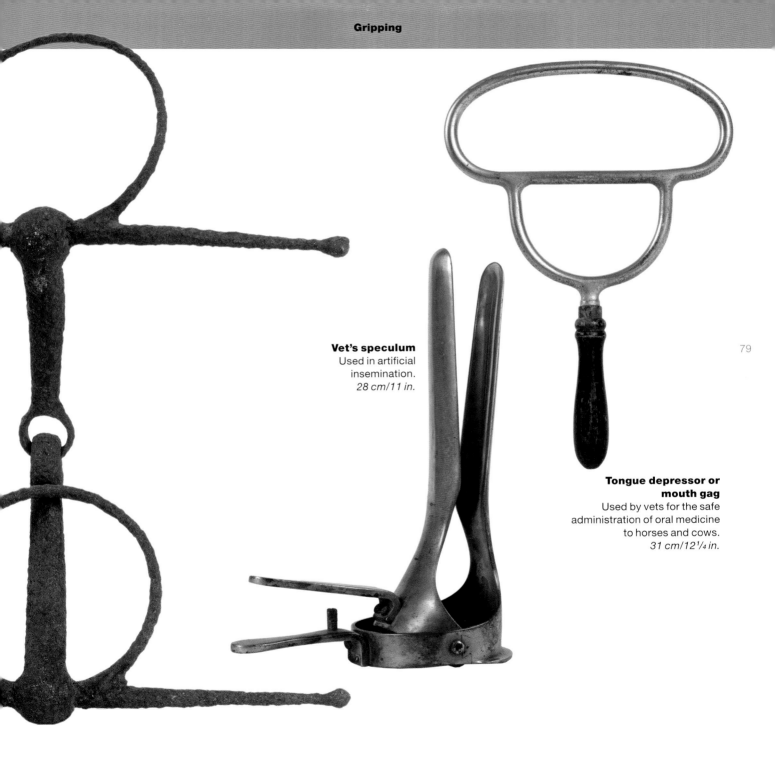

Vet's speculum
Used in artificial insemination.
28 cm/11 in.

Tongue depressor or mouth gag
Used by vets for the safe administration of oral medicine to horses and cows.
31 cm/12¼ in.

Gripping

Vine clamp
Used by French vignerons to stop the flow of sap in a vine.
15 cm/5¾ in.

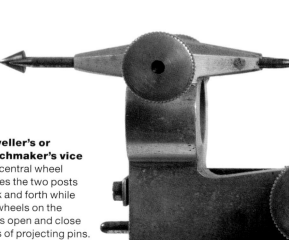

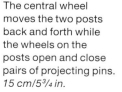

Jeweller's or watchmaker's vice
The central wheel moves the two posts back and forth while the wheels on the posts open and close pairs of projecting pins.
15 cm/5¾ in.

Gripping

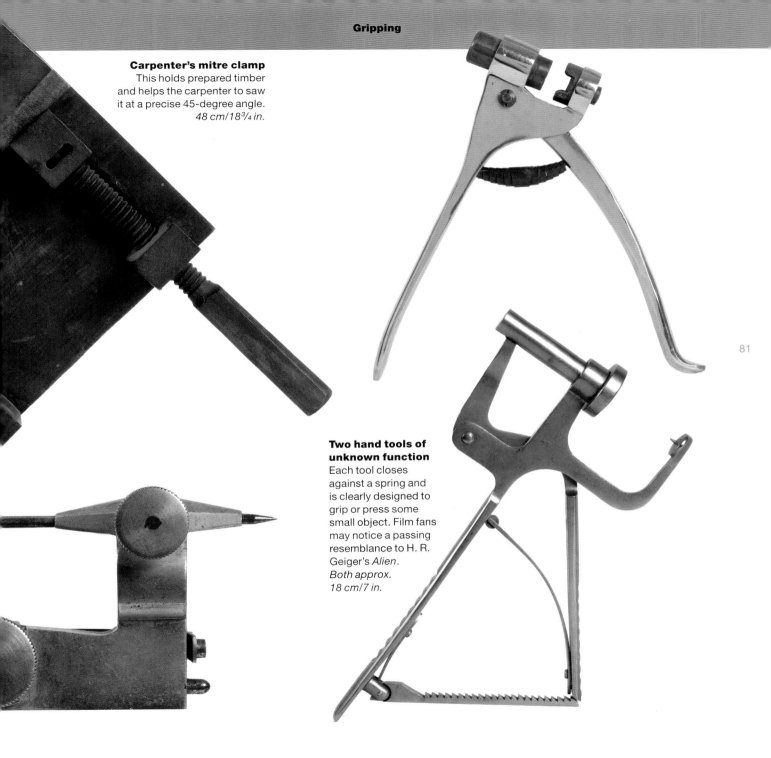

Carpenter's mitre clamp
This holds prepared timber and helps the carpenter to saw it at a precise 45-degree angle.
48 cm/18¾ in.

Two hand tools of unknown function
Each tool closes against a spring and is clearly designed to grip or press some small object. Film fans may notice a passing resemblance to H. R. Geiger's *Alien*.
Both approx. 18 cm/7 in.

Gripping

Double worm screw
Attached to the end of a set of flexible rods, this is used to screw into and pull out anything, such as dead rats or pigeons, that might be blocking a drain or a chimney.
18 cm/7 in.

Valve spring compressor
The two parts slide in and out of each other and the double-ended spanner-like piece pivots, but the function of the threaded handle is obscure.
29 cm/11½ in.

Fruit picker
When the wire stay is taken off, the central spring opens the legs and raises the wire basket towards the apple or pear. Once the fruit is wedged into the basket closing the legs against the spring gently pulls it off the branch. (Shown here upside down.)
31 cm/12¼ in.

Gripping

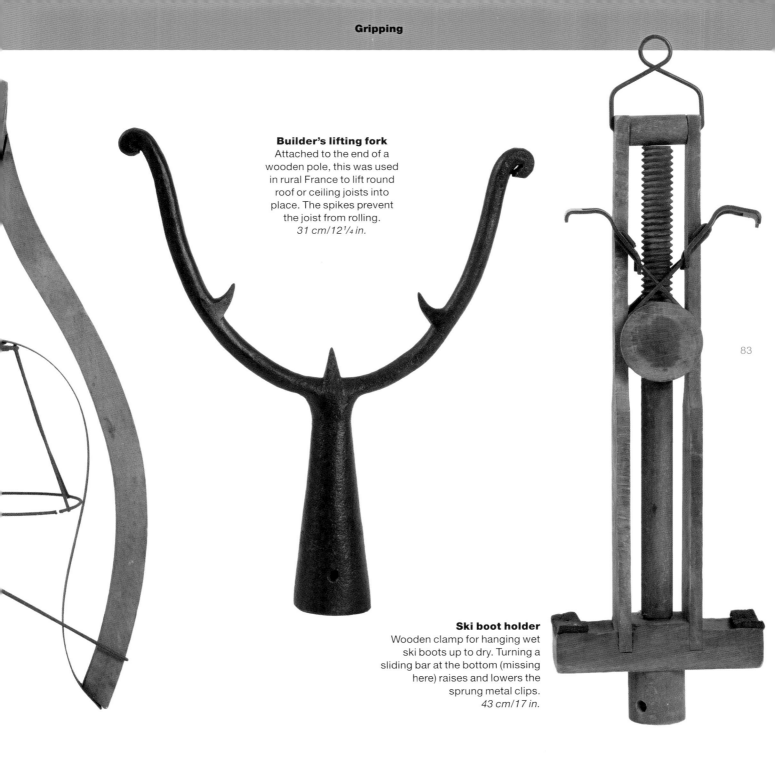

Builder's lifting fork
Attached to the end of a wooden pole, this was used in rural France to lift round roof or ceiling joists into place. The spikes prevent the joist from rolling.
31 cm/12¼ in.

Ski boot holder
Wooden clamp for hanging wet ski boots up to dry. Turning a sliding bar at the bottom (missing here) raises and lowers the sprung metal clips.
43 cm/17 in.

Gripping

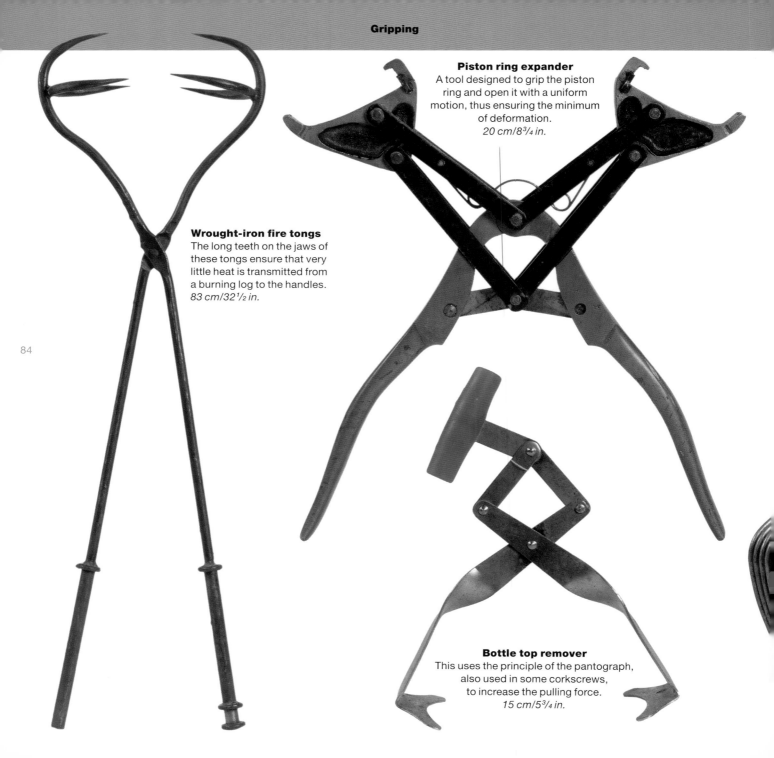

Wrought-iron fire tongs
The long teeth on the jaws of these tongs ensure that very little heat is transmitted from a burning log to the handles.
83 cm/32½ in.

Piston ring expander
A tool designed to grip the piston ring and open it with a uniform motion, thus ensuring the minimum of deformation.
20 cm/8¾ in.

Bottle top remover
This uses the principle of the pantograph, also used in some corkscrews, to increase the pulling force.
15 cm/5¾ in.

Gripping

Roast meat holder
The fork-like arms of this tool allow the user to slice a cooked joint of meat by sliding the carving knife between the tines.
28 cm/11 in.

Wooden candle snuffer
Candles in inaccessible places are usually put out using a small metal cup fixed to a pole. This works by pinching rather than crushing the wick.
87 cm/34¼ in.

Needle-nosed circlip pliers
Circlips are sprung metal rings with looped ends resembling medieval torques. Their function, when placed on a grooved cylinder, is to limit movement between that and another larger cylinder.
14 cm/5½ in.

Apple roaster
A Victorian tool for roasting apples over an open fire. A fork would drop the apple as soon as the fire had softened it.
69 cm/19½ in.

Wooden cash till
Sixty years ago, every street paper-seller and shopkeeper had one of these to help scoop up the correct change.
40 cm/15¾ in.

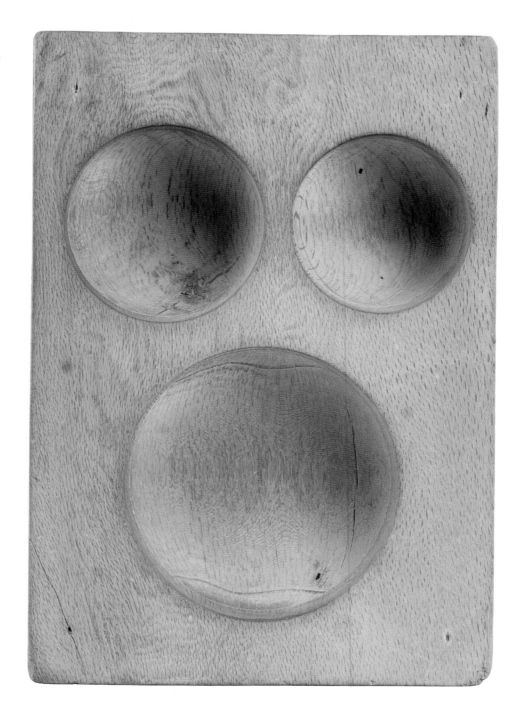

Holding
To carry, prop, trap, catch, contain, cup, cradle, enclose or support

Iron well-hook
When found in a French market, this was attached to a short length of thick wire, which suggests an agricultural rather than a domestic use. Could it be for clearing rubbish from wells or weeds from rivers? Long before Uri Geller, Picasso and the Surrealists were experimenting with the expressive possibilities of deformed forks.
24 cm/9½ in.

Our nomadic ancestors carried as little as possible, just the tools needed to kill and dismember their prey. Archaeological evidence of pottery vessels for holding water, oil or grain implies a more settled existence. A chapter entitled 'Holding' logically follows 'Hitting', 'Cutting' and 'Gripping' in an imaginary sequence of human development.

None of these objects belongs to prehistory, but they testify to the persistence of this basic function. There are containers for water, milk, air, soup, oil, salad and loose change as well as supports for necks, buttocks, noses, overcoats and wedding cakes. Some might qualify for a place in the 'Gripping' chapter, but the difference between an open hand and a clenched fist suggests a division into two groups.

Are there any conclusions to be drawn about shape? Containers of air and liquid tend to be spherical. Supports that touch the human body are usually shaped for comfort. Objects whose function is to stand in for the body, such as crutches or coat hangers, mimic the limbs they understudy. None of this is surprising. It is odd, though, that the most attractively sinuous shapes are props for pipes and bottles.

African headrest
The African equivalent of a feather pillow. It is difficult to imagine how it would be possible to sleep under these conditions, but wooden headrests were standard domestic equipment in Ancient Egypt, medieval Japan and the South Sea Islands.
19 cm/7½ in.

Holding

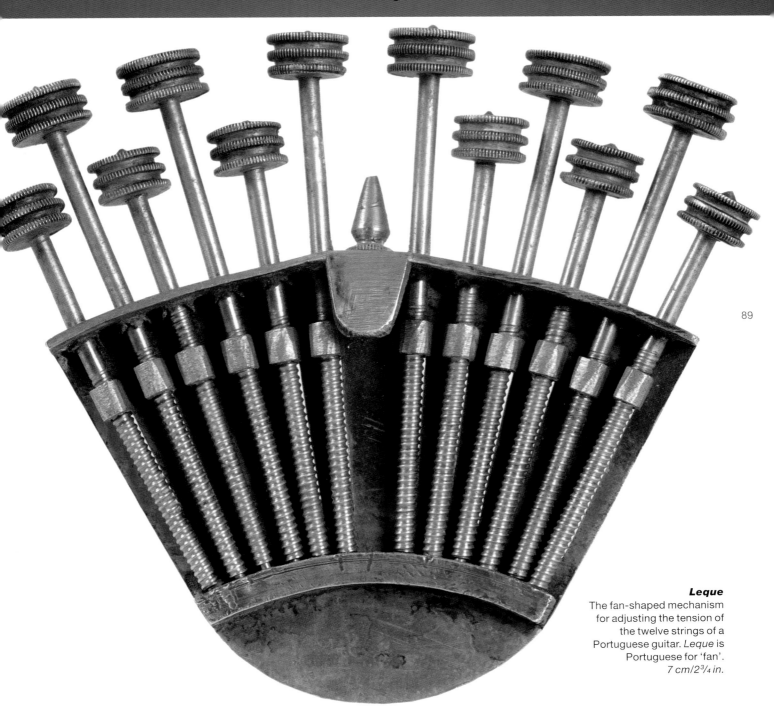

Leque
The fan-shaped mechanism for adjusting the tension of the twelve strings of a Portuguese guitar. *Leque* is Portuguese for 'fan'.
7 cm/2¾ in.

Holding

Wooden rowing seat
Ergonomically shaped for the comfort of the oarsman's buttocks, the seat rolls back and forth on metal rails to allow the longer and mechanically more efficient stroke made possible by positioning the oar's pivot point on outriggers. The holes allow water to drain from the wooden hollows.
36 cm/14¼ in.

Racing bicycle saddle
The central ventilation slit is designed to prevent the scrotal numbness that results from excessive friction.
25 cm/9¾ in.

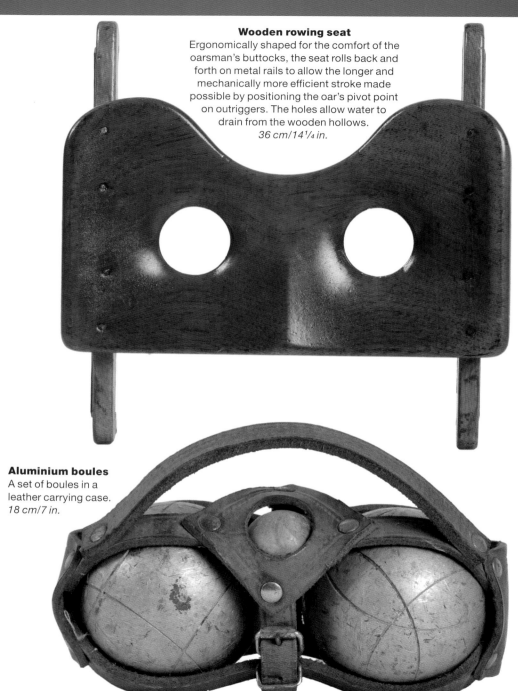

Aluminium boules
A set of boules in a leather carrying case.
18 cm/7 in.

Holding

Ceramic watercolour palette
A standard item in art shops, palettes are specially adapted plates for mixing and holding paint. The sharp corners of some well shapes help to sqeeze water from the brush.
15 cm/5 ¾ in. diameter

Two oil-painting palettes
The thumb-friendly hole allows the palette to be held comfortably by whichever hand is not wielding the brush.
*53 cm and 20 cm/
20 ¾ in. and 7 ¾ in.*

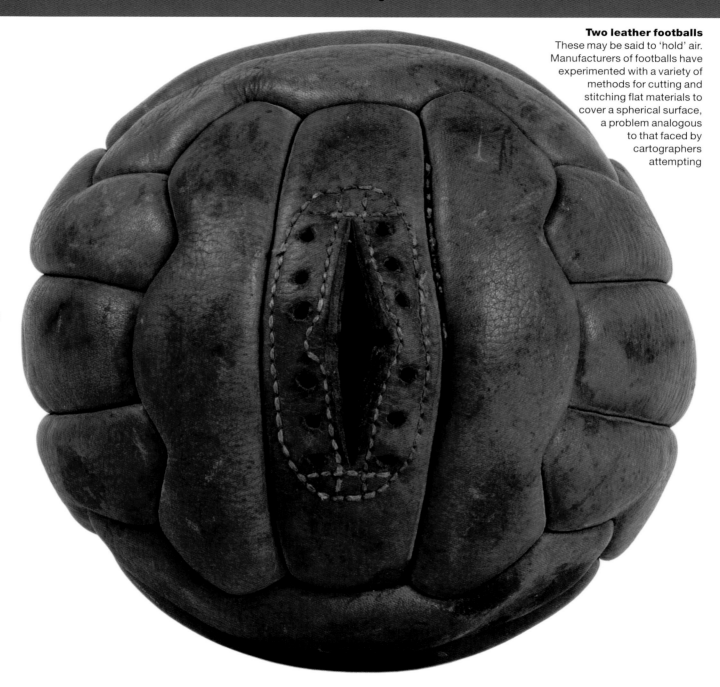

Two leather footballs
These may be said to 'hold' air. Manufacturers of footballs have experimented with a variety of methods for cutting and stitching flat materials to cover a spherical surface, a problem analogous to that faced by cartographers attempting

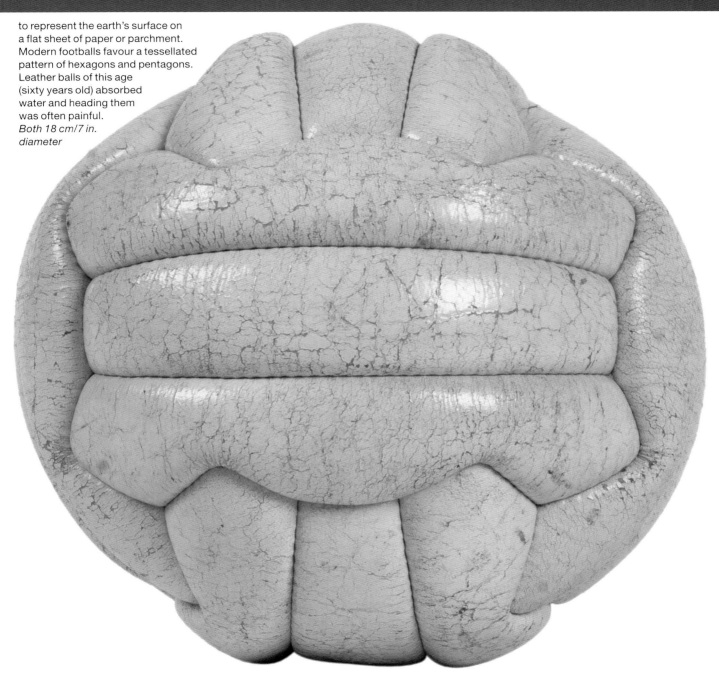

to represent the earth's surface on a flat sheet of paper or parchment. Modern footballs favour a tessellated pattern of hexagons and pentagons. Leather balls of this age (sixty years old) absorbed water and heading them was often painful.
Both 18 cm/7 in. diameter

Holding

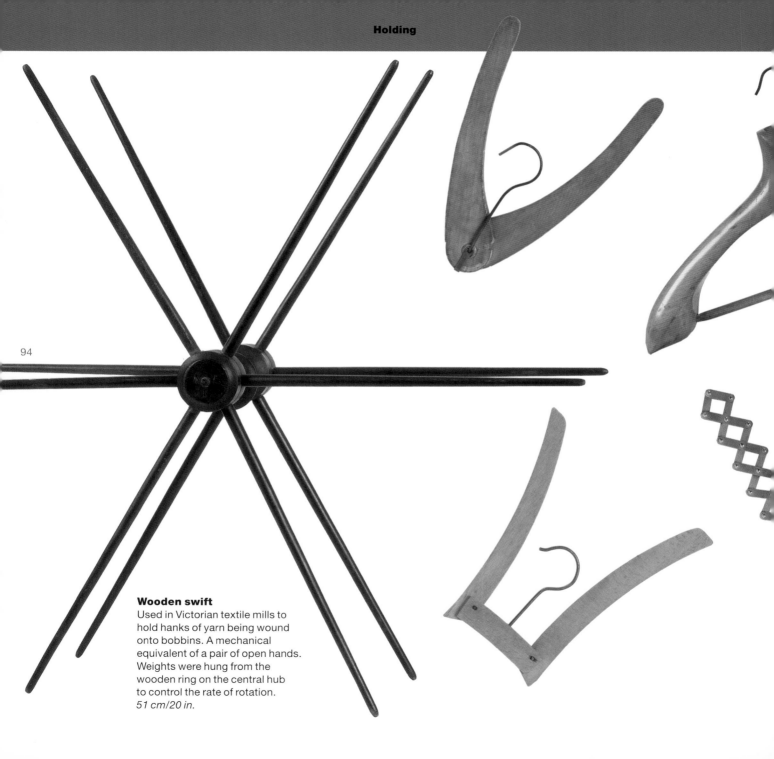

Wooden swift
Used in Victorian textile mills to hold hanks of yarn being wound onto bobbins. A mechanical equivalent of a pair of open hands. Weights were hung from the wooden ring on the central hub to control the rate of rotation.
51 cm/20 in.

Holding

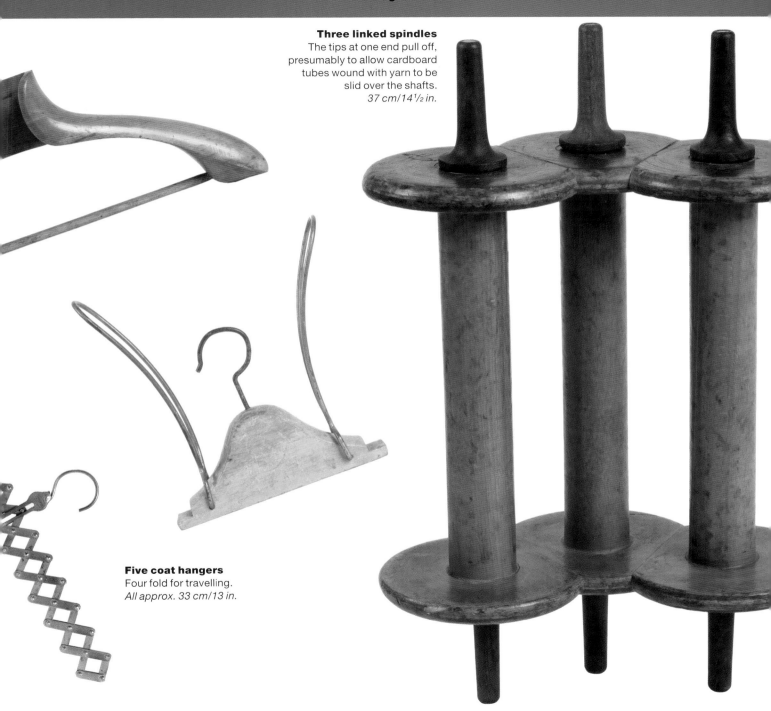

Three linked spindles
The tips at one end pull off, presumably to allow cardboard tubes wound with yarn to be slid over the shafts.
37 cm/14½ in.

Five coat hangers
Four fold for travelling.
All approx. 33 cm/13 in.

Holding

Pipe welding clamp
Found in a flea market in Paris. This allows welders to hold two lengths of pipe at any angle prior to welding.
23 cm/9 in.

Two African headrests
Above: This is little more than a wooden plate. Perhaps it was also used for eating and sitting on the ground.
19 cm/7½ in.
Right: Made for sale to tourists and carefully aged. The curved top resembles the chin rests attached to violins.
17 cm/6¾ in.

Egg holder
For lowering eggs into, and lifting them out of, boiling water. A tool for cooks too grand or busy to use a slotted spoon.
11 cm/4¼ in.

Holding

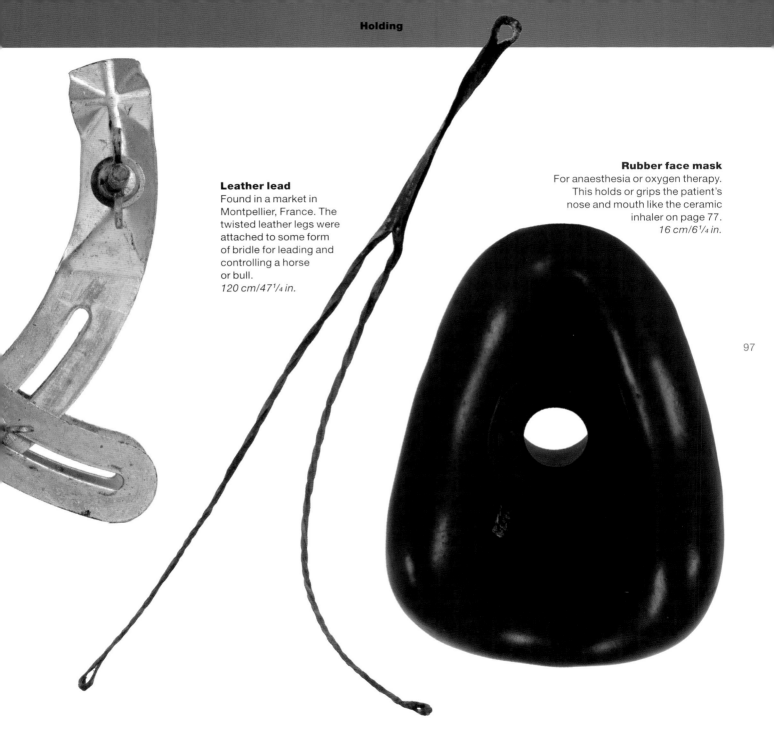

Leather lead
Found in a market in Montpellier, France. The twisted leather legs were attached to some form of bridle for leading and controlling a horse or bull.
120 cm/47¼ in.

Rubber face mask
For anaesthesia or oxygen therapy. This holds or grips the patient's nose and mouth like the ceramic inhaler on page 77.
16 cm/6¼ in.

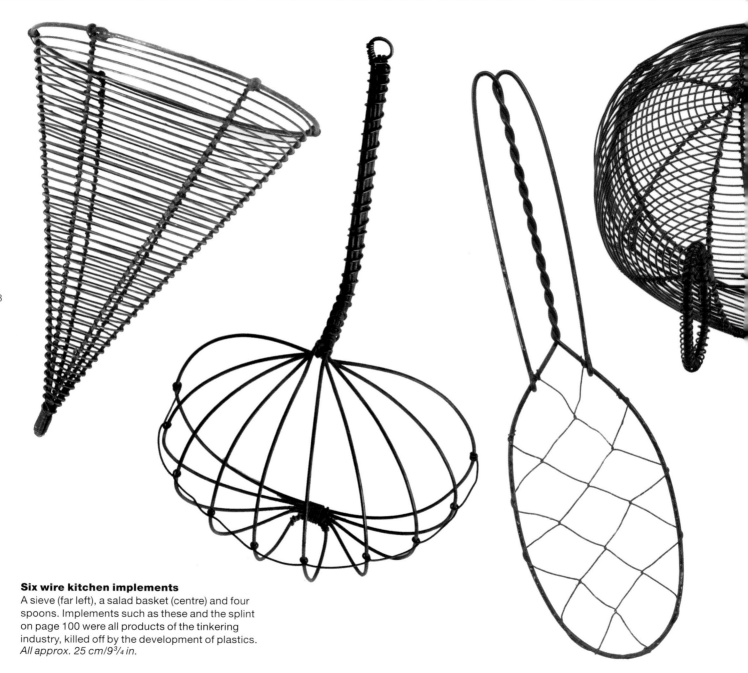

Six wire kitchen implements
A sieve (far left), a salad basket (centre) and four spoons. Implements such as these and the splint on page 100 were all products of the tinkering industry, killed off by the development of plastics. *All approx. 25 cm/9¾ in.*

Holding

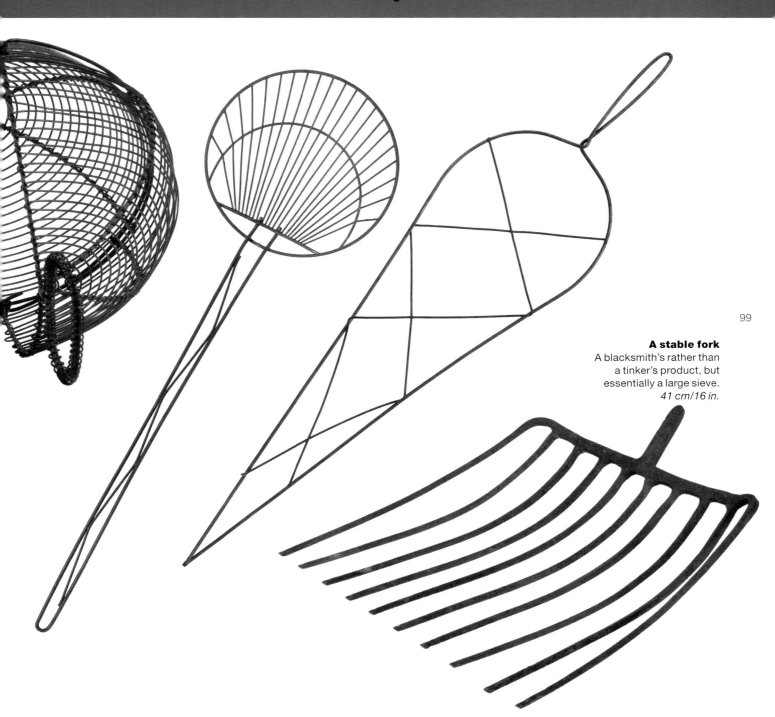

A stable fork
A blacksmith's rather than a tinker's product, but essentially a large sieve.
41 cm/16 in.

Holding

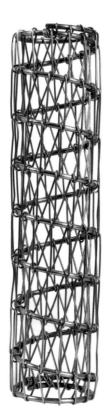

Wooden bed-stick
Bought in France from a stall selling old linen, so assumed to be a laundress's tool for lifting and smoothing sheets.
89 cm/35 in.

Wire mesh leg splint
Possibly used as an armature for a plaster cast, or as a substitute. These splints are relics of the First World War. A wire mesh arm splint appears in one of Man Ray's photographs of Lee Miller.
72 cm/28¼ in.

Steel bed-spring
The spiral spring is stiffened by a series of flexible rods.
12 cm/4¾ in.

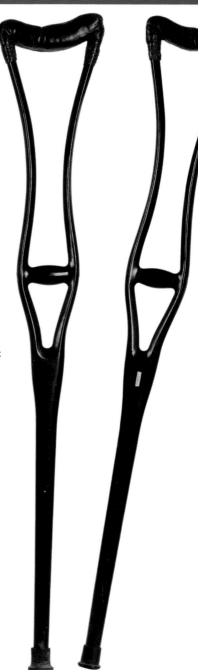

Holding

Two pairs of wooden crutches
The pair on the right are unusual in their construction and in their design. Instead of tapering to a point like the carved pair on the left, the four straight slats which form the uprights of each crutch splay out to a curved and studded 'foot'.
130 cm/51¼ in.

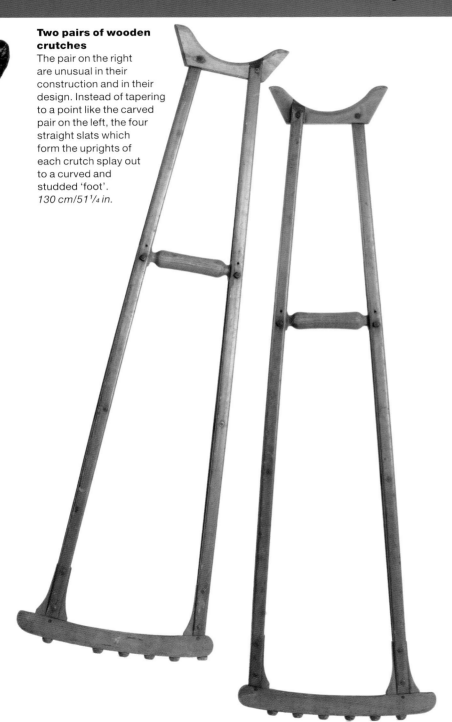

Fish trap
From the Philippines. Hand-woven in fine cane.
73 cm/28¾ in.

Wooden rack
This could be either a rack for plates or some kind of knitting aid.
34 cm/13¼ in.

Holding

Two silver-plated carving stands
For holding joints of roast meat steady.
19 cm/7½ in.

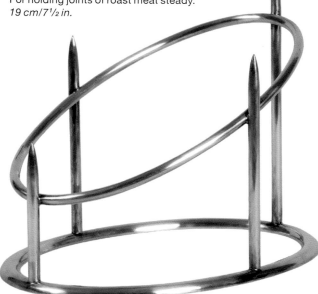

Wine bottle holder
This holds the bottle securely in a semi-recumbent position ready for serving so that sediment is not disturbed during pouring. A metal version of a Thonet rocking chair?
28 cm/11 in.

Silver-plated cake stand
Used to separate the levels of multi-tiered wedding cakes.
22 cm/8¾ in.

Holding

Laboratory retort
Glass retort with brass stand.
31 cm/12¼ in.

Tin oil can
A cross between a jug and a funnel. The carefully designed relationship between spout, mouth and handle ensures that the rate of flow is constant and not proportional to the angle at which the container is held.
22 cm/8¾ in.

African milk pot
Made from a hollow gourd.
24 cm/9½ in.

Holding

Hot-water bottle
Ceramic hot-water bottles were common before the introduction of rubber. They were heavier and could break if dropped but they could be cleaned more easily and retained heat longer. They were presumably removed from beds before occupation to avoid toe-stubbing.
30 cm/11¾ in. diameter

Rubber iced-water bag
For patients suffering from fever or a hangover.
29 cm/11½ in. diameter

Ceramic invalid feeder
The main container is protected from any surface by a hollow skirt. A flexible handle of cane or rope was presumably attached to the two ceramic hooks, allowing the feeder to be tipped into the patient's mouth.
31 cm/12¼ in.

Holding

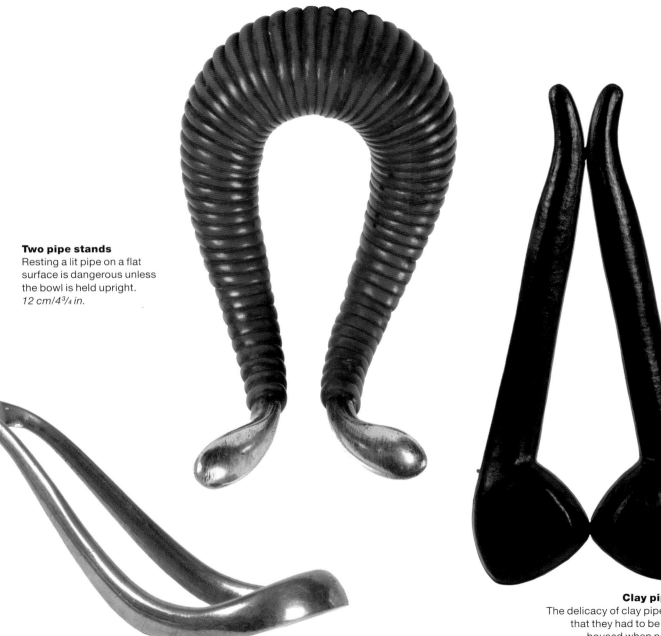

Two pipe stands
Resting a lit pipe on a flat surface is dangerous unless the bowl is held upright.
12 cm/4¾ in.

Clay pipe case
The delicacy of clay pipes meant that they had to be carefully housed when not in use.
27 cm/10½ in.

Stainless-steel blade for kitchen waste grinder
This device, popular in the 1950s, replaced the standard trap on a kitchen sink with a grinder advertised as being capable of pulverizing and reducing to slurry any kitchen waste, thus solving the problem of waste disposal in tall buildings. This proved irresistible to small boys eager to test its strength by inserting spoons, glass, fingers, etc.
20 cm/7¾ in. diameter

Rubbing
To roll, mix, comb, crush, grind, mash, scrape, massage, mince or whisk

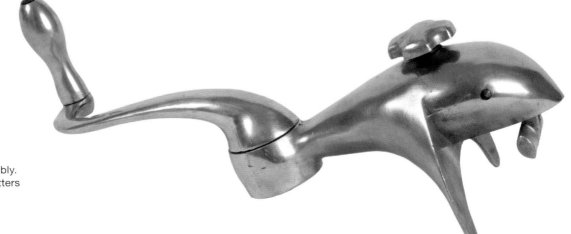

Aluminium clamp for a domestic food grinder
This is only half of the assembly. The food funnel, grinding cutters and spout are missing.

Rubbing

In the history of mankind's development, tools for rubbing and grinding were presumably needed once the nomadic stage was replaced by a more settled life that allowed for the growing of crops. It is interesting that two of the oldest objects in this collection are the stone pestles (page 111) from Hawaii. These were used to crush breadfruit, which the natives call *poi*.

With a few exceptions, all the tools illustrated are related either to food (kitchen whisks, pasta cutters), flesh (combs, massage rollers) or art (paint scrapers, sculptor's rasps). There are many other substances that are rolled, crushed or scraped but they seem not to have involved tools which caught the collector's eye.

Readers alert to visual clues may find here references to *Moby Dick*, Giacometti's sculptures, flying saucers, golf, Leonardo da Vinci and Miss Piggy from the Muppets.

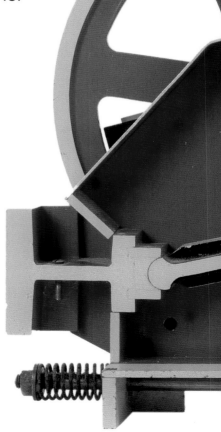

Sectional model of an industrial stone-crushing plant
A salesman's demonstration model with moving parts operated by the flywheel on the left.
44 cm/17¼ in.

Rubbing

109

Rubbing

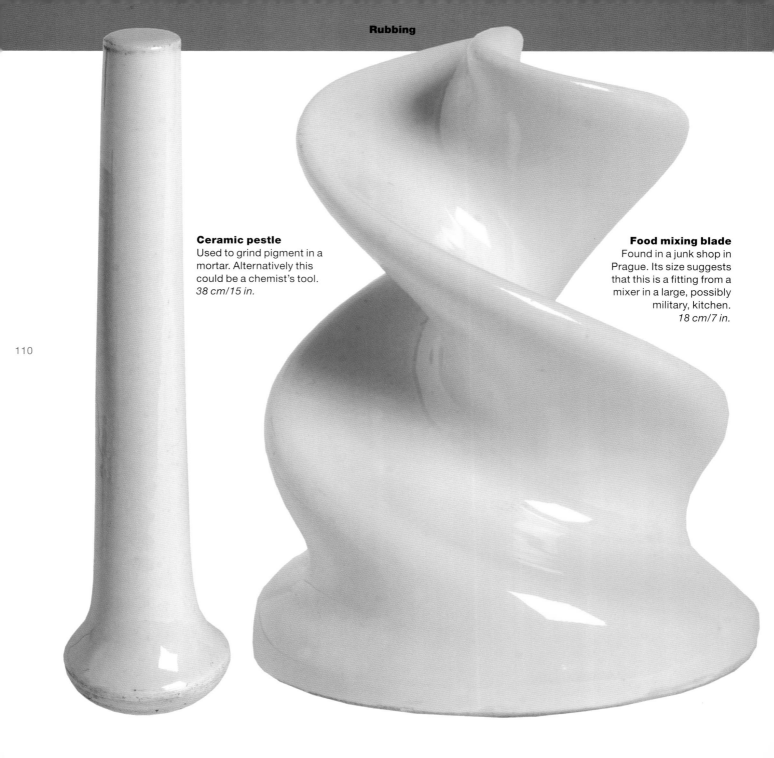

Ceramic pestle
Used to grind pigment in a mortar. Alternatively this could be a chemist's tool.
38 cm/15 in.

Food mixing blade
Found in a junk shop in Prague. Its size suggests that this is a fitting from a mixer in a large, possibly military, kitchen.
18 cm/7 in.

Rubbing

Pastry roller
The advantage of this tool is that it can be used with one hand. Other methods of thinning pastry – the rolling pin or the frisbee spin used by pizza cooks – usually require two.
25 cm/9¾ in.

Two stone breadfruit pounders
Stone pounders were used in the Fijian Islands to mash fermented foodstuff and starchy roots. The resulting mash was sometimes spiced with human liver.
Both approx. 21 cm/8¼ in.

Rubbing

Four massage rollers
To judge by the regularity with which these turn up at antique fairs,
massage seems to have been a popular pastime in the 1920s and 1930s.
Presumably they were self-administered as slimming aids rather than as devices
to give sensual pleasure. The all-wooden roller at the bottom is for massaging the feet.
All approx. 36 cm/14¼ in.

Rubbing

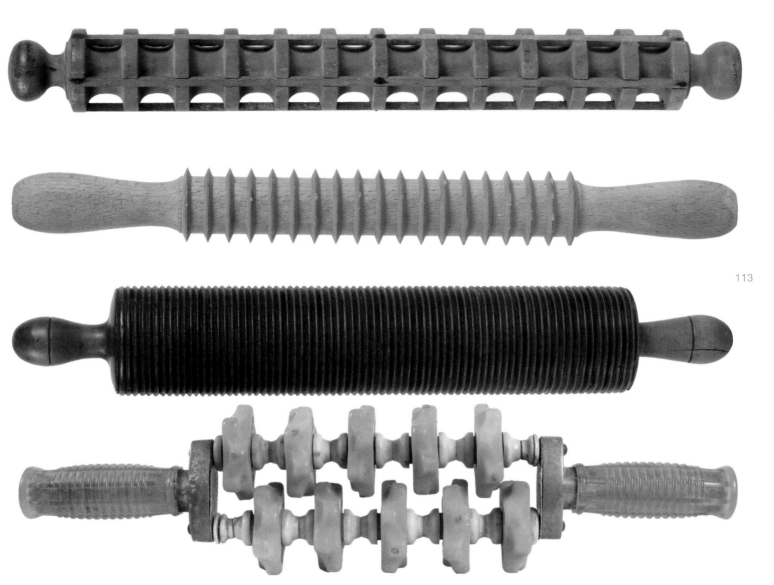

Four kitchen rollers
The top two are for cutting pasta. The large-diameter grooved roller second from the bottom is for crushing oats. The double roller at the bottom is for putting dimples into pastry.
All approx. 46 cm/18 in.

Rubbing

Paint mixing bit
When attached to an electric drill, this tool stirs and mixes paint.
34 cm/13¼ in.

Two sculptor's tools
A metal gouge for scraping and a wooden spatula for pressing the soft clay on a model.
Both 18 cm/7 in.

Film developing tank
This was presumably a piece of equipment for photographers who needed to work outdoors using portable darkrooms. Developing fluid can be poured into the tank through the funnel next to the lid and drained at the bottom.
26 cm/10¼ in.

Rubbing

Four kitchen whisks
In the whisk on the far right the blades rotate when the bar is pushed up and down along the two twisted spirals. Two hands would be needed. The whisk next to it is more efficient in that only one hand is needed to pump the handle down the twisted spiral of the central shaft. The weighted arms add centrifugal spin to the blades.
All approx. 24 cm/9½ in.

Rubbing

Five massage devices
From left to far right:
Mechanical blood circulator. The vibrating rubber cup stimulates the blood around a wound, enabling faster healing.
31 cm/12¼ in.
Wooden back stick. An orthopaedic device for correcting posture in children.
78 cm/30¾ in.
Two-handled spherical wooden roller.
30 cm /11¾ in.
Double-ended ivory and wood face massager.
16 cm/6¼ in.

Rubbing

Plastic multi-ball massager.
16 cm/6¼ in.

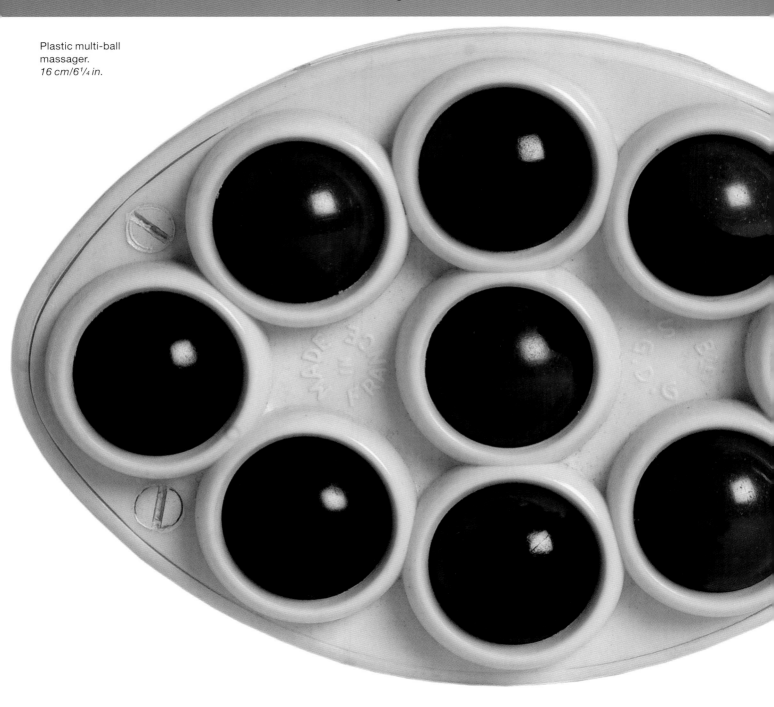

Rubbing

Coconut chopping board
The shaft of the blade, hinged at one end, is used to split the nuts. The serrated tip is then used to scrape out the flesh.
35 cm/13¾ in.

Corn cob stripper
Allegedly for stripping corn off cobs, presumably after they have been cooked. The cob is rotated in the central cleft and the kernels collected in the surrounding dish.
20 cm/7¾ in.

Rubbing

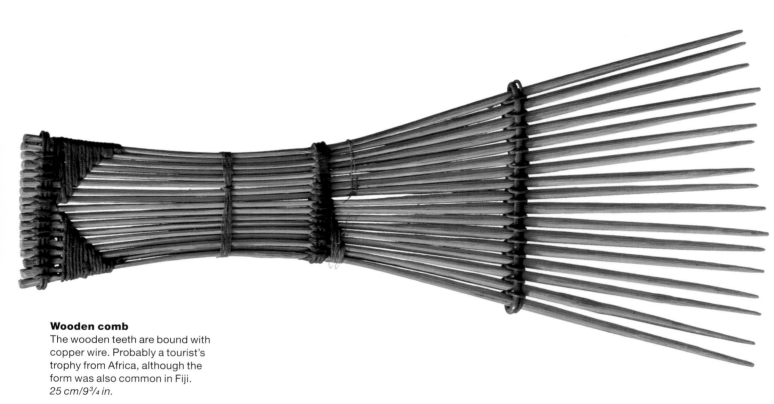

Wooden comb
The wooden teeth are bound with copper wire. Probably a tourist's trophy from Africa, although the form was also common in Fiji.
25 cm/9¾ in.

Rubbing

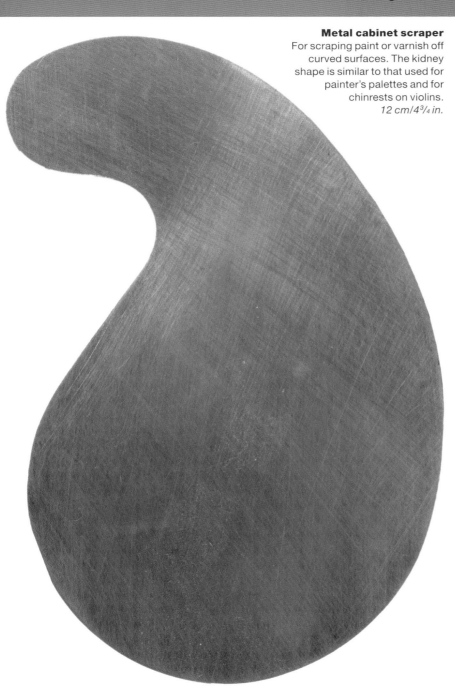

Metal cabinet scraper
For scraping paint or varnish off curved surfaces. The kidney shape is similar to that used for painter's palettes and for chinrests on violins.
12 cm/4¾ in.

Hackle board
Essentially a primitive comb made of metal with a wooden base. Used to straighten flax fibres prior to spinning.
40 cm/15¾ in.

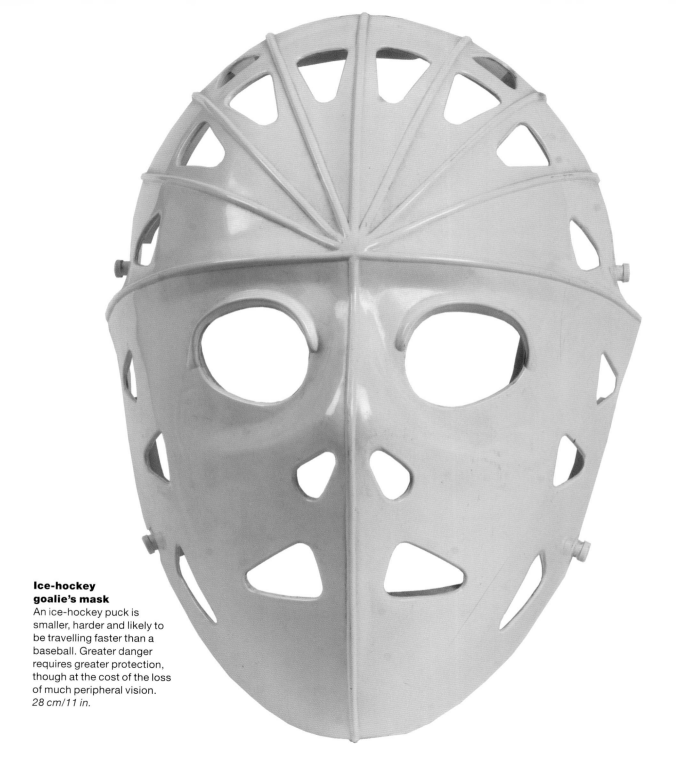

Ice-hockey goalie's mask
An ice-hockey puck is smaller, harder and likely to be travelling faster than a baseball. Greater danger requires greater protection, though at the cost of the loss of much peripheral vision.
28 cm/11 in.

Shielding
To guard, protect, cover, screen, hide, deflect, insulate, mask or defend

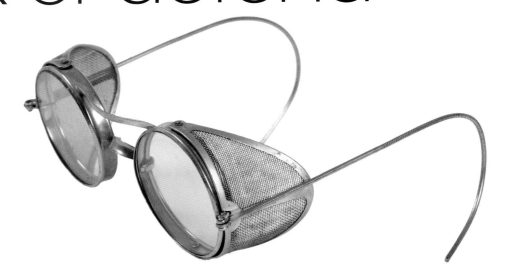

Motorist's or cyclist's protective spectacles
The mesh of the hinged side pieces protects the wearer from the debris thrown up from unsurfaced roads.
18 cm/7 in.

Shielding

In the summer of 1995, the Design Department of the Museum of Modern Art in New York mounted a small exhibition of masks. These were not the tribal masks admired and collected by the pioneers of Modernism, but working masks whose function was to protect their wearers from welding sparks, baseballs, ice-hockey pucks, burning oil wells, angry bears or swarming bees.

Partly because of their anonymity, these masks had an astonishing presence. They were anonymous both in the sense that they conferred anonymity on their wearers and in the sense that they were not the work of the famous designers who were the usual subjects of the department's exhibitions.

Masks fascinate us because of what they conceal as they protect. The masked can see us but we cannot see them. They appear aggressive because they imply that we, the unmasked, are a threat. Perhaps this is why they are the symbols by which we recognize so many fictional heroes and villains, from Batman and Zorro to Darth Vader.

Spiked plastic pineapple
Anti-burglar device threaded onto a metal pole mounted on top of a wall. The pineapples are slotted into each other but rotate freely on the pole. These are an improvement on the traditional row of smashed bottle glass set in cement, now illegal.
20 cm/7³/₄ in.

Shielding

Some masks protect both the wearer and the public. In the Middle Ages people with facial injuries or wasting diseases such as syphilis or leprosy wore masks that protected the wearer from prying eyes and the viewer from the spectacle of disfigurement. Hannibal Lecter's mask protects us from him but, by isolating his glittering eyes and ravenous, caged lips, makes his face truly terrifying.

This explains the presence, in this chapter devoted to objects that shield and protect, of three masks. There are many other types of mask for which the search continues, such as firefighter's masks like those photographed by Lee Miller in the Blitz, gunnery masks like the one found by Breton and Giacometti in a Paris market and illustrated in Breton's book *L'Amour fou,* and early welding masks. The other objects in this chapter protect people from dust, water and sunlight, buildings from burglers and dead pigeons, dining tables from hot casseroles, and ironing boards from hot irons.

Shielding

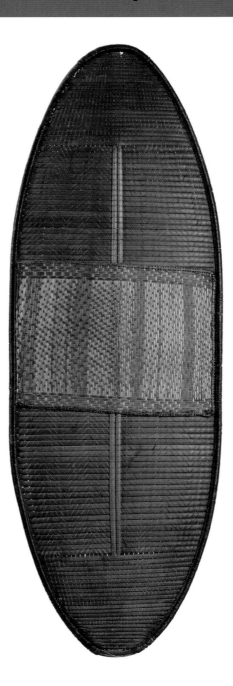

Shielding

Chinese mud boot
Made of leather covered with black lacquer. The foot size of whoever used it was minute.
37 cm/14½ in.

Plaster doll face mould When air is sucked out of the space between a mould and a thin sheet of heat-softened plastic, the plastic cools into an exact replica of the pattern. *21 cm/8¼ in.*

Moulding
To cast, fashion, bend, form, press, forge, shape, stamp or stretch

Moulding

Of the four basic ways of shaping materials – cutting, constructing, bending and moulding – the last is the most efficient in its use of materials, but expensive in the costs of production.

Imagine that you need a chair. You could carve it out of a single tree trunk but that would take a long time and waste three-quarters of your material. If you sawed the tree into lengths you could construct several chairs. They might not be very comfortable but you would have less waste and more chairs. If you had the machinery and the right kind of wood, parts of the chair could be softened with steam and bent to more comfortable shapes.

Finally, with an even greater investment in plant you could reduce the tree to sawdust, add water and a setting agent and pour the resulting soup into moulds to produce any desired shape with no wastage at all.

All man-made objects, from bicycles to bridges, can be analysed in terms of these four basic

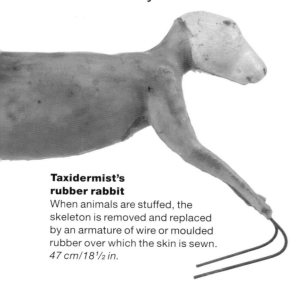

Taxidermist's rubber rabbit
When animals are stuffed, the skeleton is removed and replaced by an armature of wire or moulded rubber over which the skin is sewn. 47 cm/18½ in.

Moulding

processes, and they define the art of sculpture: cutting and carving (Michelangelo, Brancusi, Moore), construction (David Smith, Anthony Caro), bending (Richard Serra, Alexander Calder), and casting and moulding (any sculptor who wants to make multiple copies from an original).

To understand the importance of this final process, picture yourself, after a visit to the dentist, sitting on a café terrace in a plastic chair pouring tonic into a glass already half full with ice cubes. Almost everything in this scene – the chair, the terrace, the ice, the glass, the bottle and your new tooth – are the result of a casting process. They have all, at some stage, been sufficiently liquid to be given their final shape in moulds – metal moulds for the glass, chair and bottle, wooden moulds for the concrete terrace and rubber moulds for the ice and your tooth.

The objects in this chapter relate either to the forming or the casting process. They are either tools used to stretch and give shape to flexible materials such as leather and felt, or moulds into which liquids such as raw egg and molten chocolate are poured. The final objects included are *results* of the casting process: medical models of human body parts that failed to find a home in any of the earlier chapters.

Adjustable hat mould
This takes a positive copy of the shape of a customer's head recorded by the device on page 177.
18 cm/7 in.

Moulding

Wooden pattern
For a vessel used in making chocolate. A pattern is a wooden shape that is pressed into a closely packed mixture of oil and fine sand. When withdrawn, it leaves a negative space into which molten metal is poured.
77 cm/
30¼ in.

Wooden pattern
Opposite: This seems to be for the handle of a large tap or valve. The central boss would be drilled out to hold the spindle. Found in a Brussels street market.
24 cm/9½ in.

Portuguese tile mould
Above: Placed on a tile base, this allows the manufacturer to apply a complex design by pouring coloured slip (liquid clay) into the different sections of the mould. When the slip is dry the tile is glazed.
20 cm/7¾ in.

Moulding

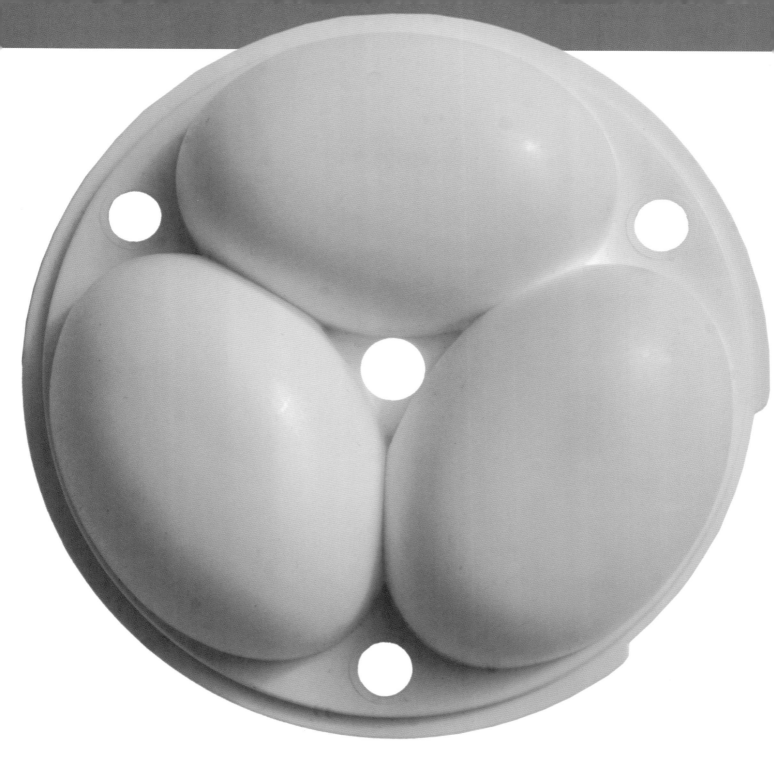

Moulding

Plastic mould for poaching eggs
Left: The product of a moulding process (vacuum-forming), which in turn moulds the eggs that it holds as they cook.
17 cm/6¾ in. diameter

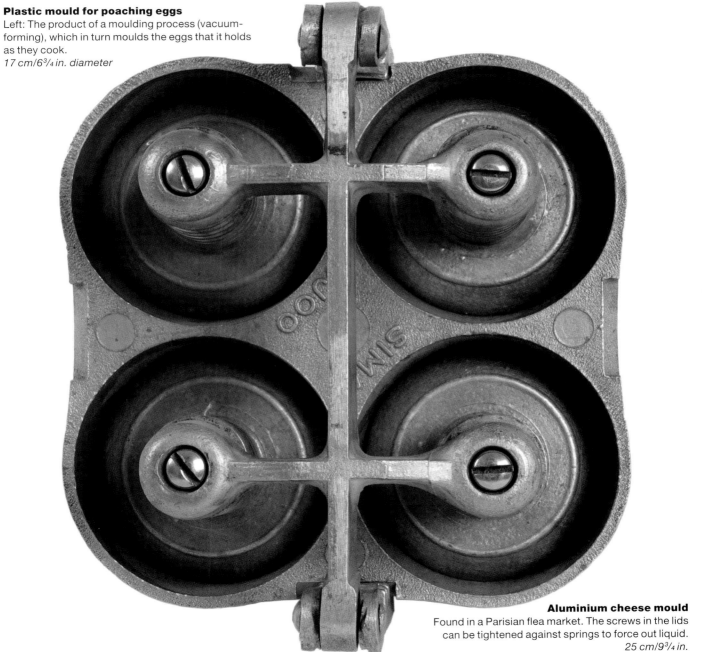

Aluminium cheese mould
Found in a Parisian flea market. The screws in the lids can be tightened against springs to force out liquid.
25 cm/9¾ in.

Moulding

Wooden casting pattern
This would have formed a quarter of a circular iron collar.
26 cm/10¼ in.

Pharmacist's mould
For casting suppositories. When the suppositories have set they can be removed by loosening the two screws and separating the three parts of the mould.
15 cm/5¾ in.

The Cresco patented trouser creaser
This creases trousers by stretching them from the inside of the leg rather than pressing the outside. It is ingeniously designed to adapt to the varying width of the trouser leg.
87 cm/34¼ in.

Moulding

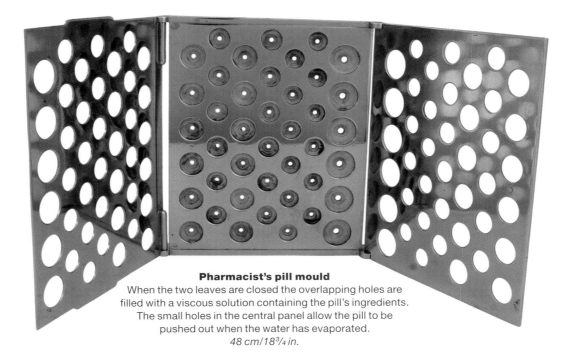

Pharmacist's pill mould
When the two leaves are closed the overlapping holes are filled with a viscous solution containing the pill's ingredients. The small holes in the central panel allow the pill to be pushed out when the water has evaporated.
48 cm/18¾ in.

African casting pattern
For casting ribbed metal plates that are then bent and hammered into bracelets.
30 cm/11¾ in.

Moulding

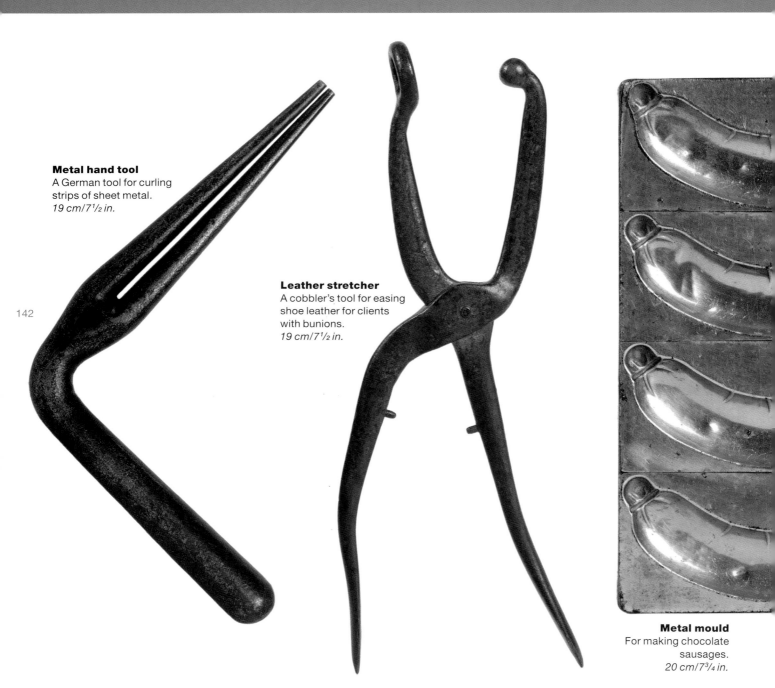

Metal hand tool
A German tool for curling strips of sheet metal.
19 cm/7½ in.

Leather stretcher
A cobbler's tool for easing shoe leather for clients with bunions.
19 cm/7½ in.

Metal mould
For making chocolate sausages.
20 cm/7¾ in.

Moulding

Conical wooden former
Used in conjunction with a wooden anvil to shape the sheet-metal tips of organ pipes.
33 cm/13 in.

Glass cucumber forcer
Cucumbers grown with the help of these devices were easier to pack and to sell.
42 cm/16½ in.

Neck-tie stretcher
The width of the stretcher can be increased by twisting the wooden handle.
53 cm/20¾ in.

Moulding

Round wooden patterns
Left and far right: For casting the metal cams that control the movement of the spindles over which carpet passes as it is being manufactured.
Both 58 cm/22¾ in. diameter

Wooden patterns
Right: For machine parts.
Both 70 cm/27½ in.

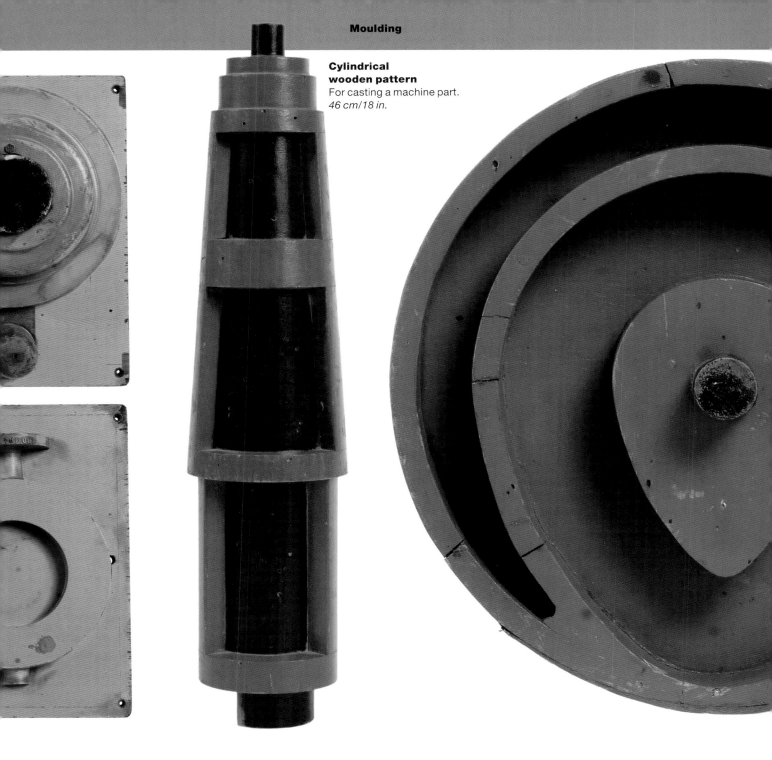

Moulding

Cylindrical wooden pattern
For casting a machine part.
46 cm/18 in.

Moulding

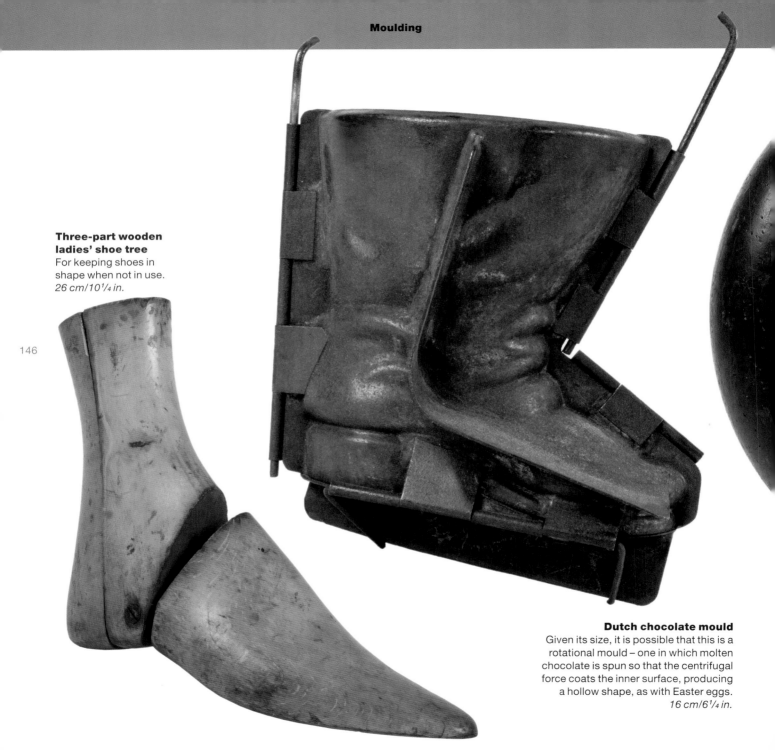

Three-part wooden ladies' shoe tree
For keeping shoes in shape when not in use.
26 cm/10¼ in.

Dutch chocolate mould
Given its size, it is possible that this is a rotational mould – one in which molten chocolate is spun so that the centrifugal force coats the inner surface, producing a hollow shape, as with Easter eggs.
16 cm/6¼ in.

Moulding

Wooden boot last
The extraordinary width at the toe suggests that this may have been used for making boots for deep-sea divers.
29 cm/11 ½ in.

Wooden hatter's former
Used in conjunction with crown moulds to form the curved brims of ladies' hats.
35 cm/13 ¾ in. diameter

Moulding

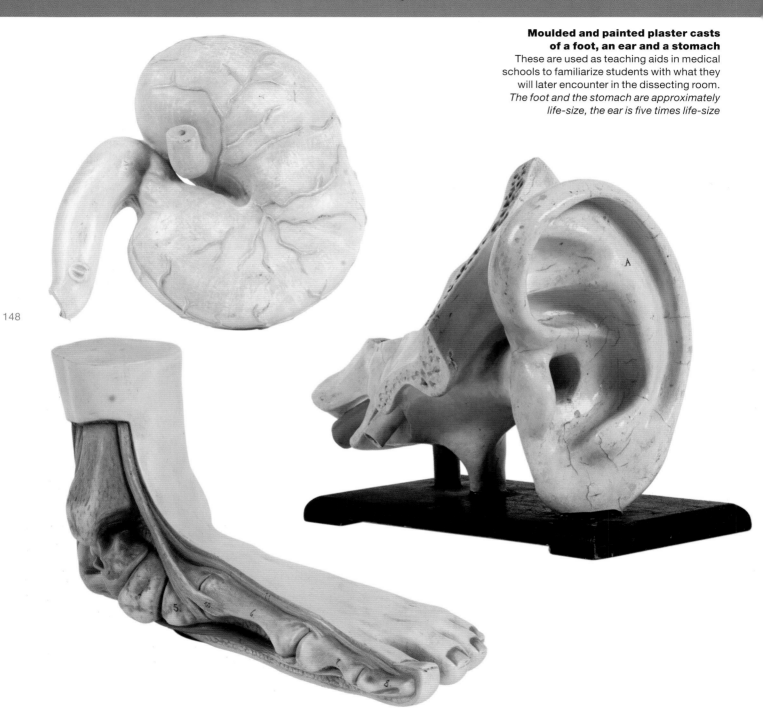

Moulded and painted plaster casts of a foot, an ear and a stomach
These are used as teaching aids in medical schools to familiarize students with what they will later encounter in the dissecting room.
The foot and the stomach are approximately life-size, the ear is five times life-size

Moulding

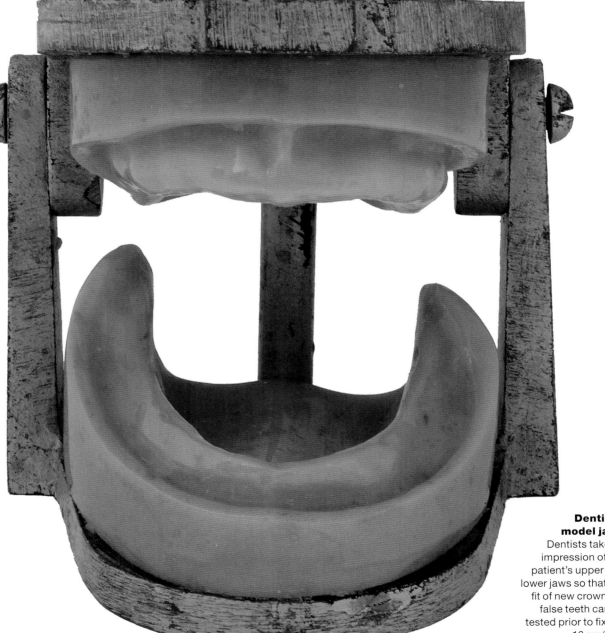

Dentist's model jaws
Dentists take an impression of the patient's upper and lower jaws so that the fit of new crowns or false teeth can be tested prior to fixing.
13 cm/5 in.

Two car window springs
When car windows are wound down with a handle the springs are tightened so that their force helps the window to close when the handle is wound up.
10 cm/4 in. diameter

Spreading
To expand, diffuse, disperse, circulate, dilate, open, scatter or transmit

Gramophone spring
When wound, the spring stores energy which it then transmits to the turntable. The word 'gramophone', like the thing it denotes, is on the brink of extinction.
30 cm/11 3/4 in. diameter

Spreading

Although there is a slightly miscellaneous aspect to this chapter, it is interesting that an elastic interpretation of the word 'spreading', which stretches from gramophone horns to radiators, should produce such a range of expressive shapes. Why is this? Probably because, if one excludes hats and flesh, the things spread – heat, sound, energy, gas, dust, light and water – are formless. The spreading devices, therefore, often mimic plants whose survival depends on their ability to grab as much light and moisture as possible in a confined space, or animals that regulate their body heat by expanding or contracting their surface area.

Once again, a debt to the world of art must be acknowledged. The visual appeal of musical instruments, especially the guitar, the tuba and the violin, is evident in the paintings of Picasso and René Magritte. Gramophone horns have, at times, obsessed Magritte, Caro and Kapoor. Le Corbusier displayed a gas ring alongside his collection of simple pottery and other found objects. Both Picasso and Joan Miró used gas rings as humorous elements in their sculpture. And if the early Modernists had not admired and collected African nail fetishes, the flared cylinder head on page 158 might have gone unnoticed.

Sugar shaker
This is a standard shape for a salt cellar. Sugar usually appears at the table in a bowl, as lumps or in a sachet.
15 cm/5¾ in.

Spreading

Gas ring
The size and shape of the jets suggest that this comes from a water heater rather than a domestic stove.
17 cm/6¾ in. diameter

Spreading

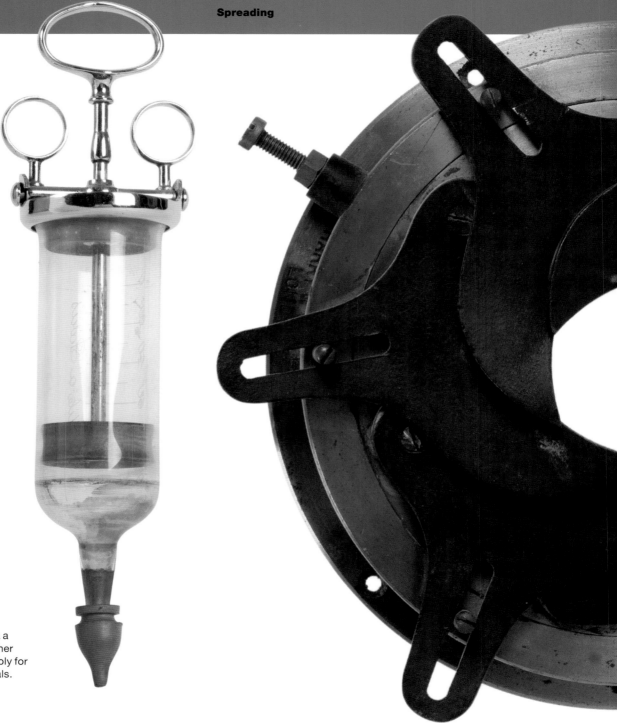

Vet's large glass syringe
This injects liquid via a small rubber teat rather than a needle. Possibly for feeding young animals. *21 cm/8¼ in.*

Spreading

Brass iris diaphragm
Early radiotherapy device for focussing radiation on patients. This operates like the iris shutter of a camera but is built like a ship's porthole.
23 cm/9 in. diameter

Wooden dust pump
A large wooden syringe for blowing dust out of the inaccessible corners of radiators, organs or pianos. An invaluable tool for collectors too lazy to wipe the dust from the objects on their shelves.
82 cm/32¼ in.

Spreading

Gramophone horn
The classic bell-shaped gramophone horn is illustrated on page 165. This double horn may be an early attempt at stereo, like the double motor horn, also on page 165.
40 cm/15¾ in.

Vet's syringe
The pistol grip action resembles that on sealant guns.
25 cm/9¾ in.

Spreading

157

Spreading

Two aluminium motorcycle cylinder heads
The fins serve to dissipate the heat of the engine. The fins on some species of lizard serve the same function.
Both 24 cm/9½ in. diameter

Spreading

Spreading

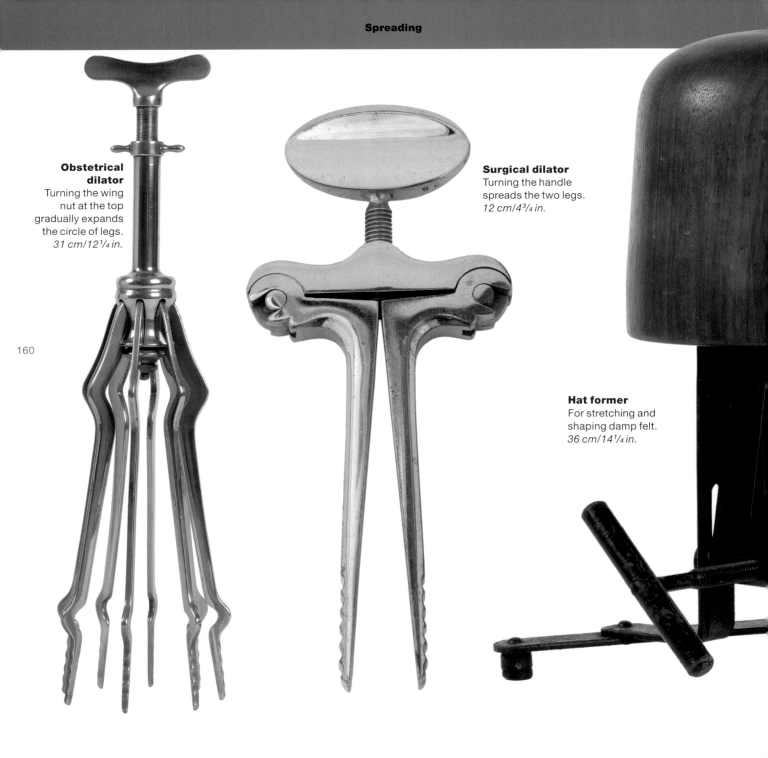

Obstetrical dilator
Turning the wing nut at the top gradually expands the circle of legs.
31 cm/12¼ in.

Surgical dilator
Turning the handle spreads the two legs.
12 cm/4¾ in.

Hat former
For stretching and shaping damp felt.
36 cm/14¼ in.

Spreading

Dairyman's teat stretcher
If the cups on a milking machine were too tight for a cow's teat this device was used to stretch them.
16 cm/6¼ in.

Plumber's pipe stretcher
Before the change to copper, plumbers, as their name indicates, worked with lead piping, which was soft and flexible. Pipes were joined not with special fittings, but by flaring the end of one section, using this tool, so that a sleeve joint could be made with the next section. This resulted in an ugly swelling, as though the pipe had swallowed and was digesting a small animal.
25 cm/9¾ in.

Spreading

Radiator brush
The Mohican profile of the soft bristles is puzzling.
142 cm/56 in.

Brass water pump
Placed in a bucket of water, this produces a jet of liquid when the handle is pumped up and down. An early device for fighting fires where there was no mains pressure.
87 cm/34¼ in.

Fire hose jet spreader
Fitted to the nozzle of a fireman's water hose when the operator wishes to create a wide spray of water rather than a concentrated jet.
15 cm/5¾ in.

Spreading

African fly whisk
Made from ivory and horsehair, almost certainly for the tourist trade but beautiful nonetheless.
56 cm/22 in.

Japanese fan
Made with coiled and woven grass.
34 cm/13¼ in. diameter

Spreading

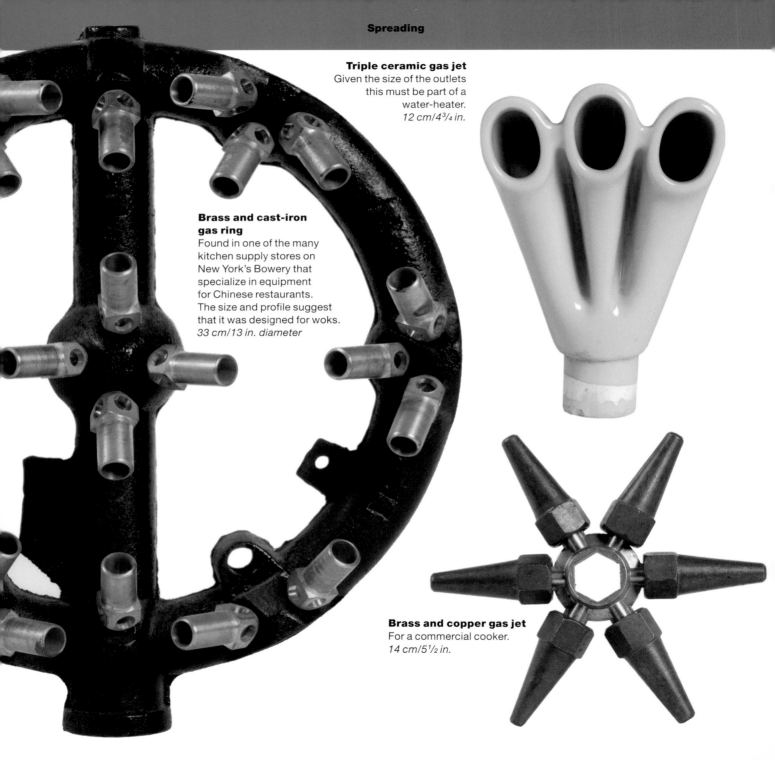

Triple ceramic gas jet
Given the size of the outlets this must be part of a water-heater.
12 cm/4¾ in.

Brass and cast-iron gas ring
Found in one of the many kitchen supply stores on New York's Bowery that specialize in equipment for Chinese restaurants. The size and profile suggest that it was designed for woks.
33 cm/13 in. diameter

Brass and copper gas jet
For a commercial cooker.
14 cm/5½ in.

Spreading

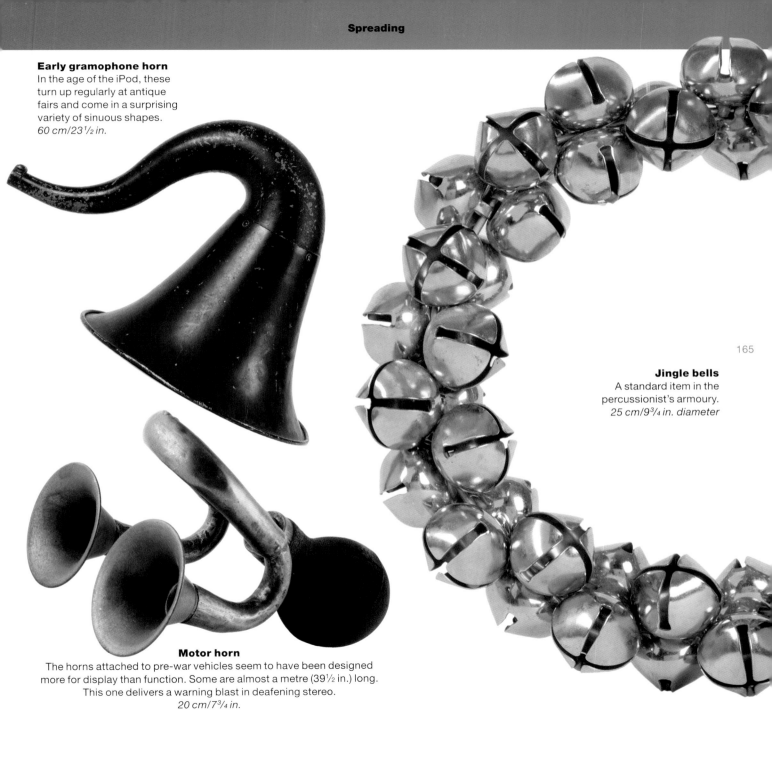

Early gramophone horn
In the age of the iPod, these turn up regularly at antique fairs and come in a surprising variety of sinuous shapes.
60 cm/23½ in.

Jingle bells
A standard item in the percussionist's armoury.
25 cm/9¾ in. diameter

Motor horn
The horns attached to pre-war vehicles seem to have been designed more for display than function. Some are almost a metre (39½ in.) long. This one delivers a warning blast in deafening stereo.
20 cm/7¾ in.

Spreading

Cornet
For spreading sound. Note the finger-friendly hook that allows this to be played with one hand.
30 cm/11¾ in.

Spreading

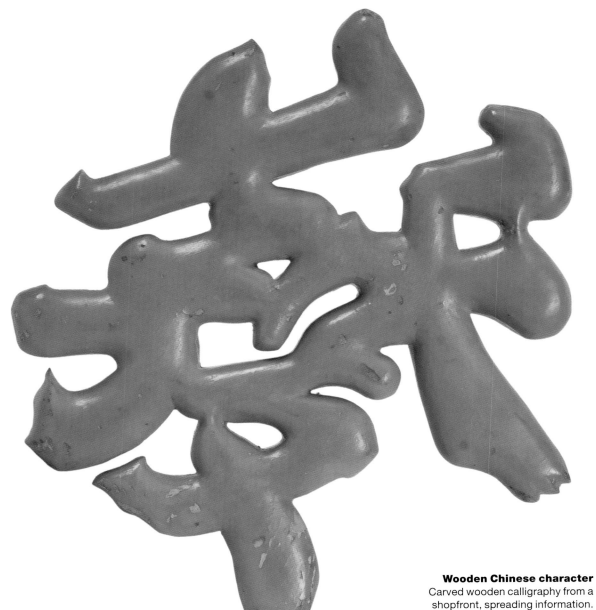

Wooden Chinese character
Carved wooden calligraphy from a shopfront, spreading information. Pronounced *fa*, it means to shoot, stand up, leave or appear.
27 cm/10½ in.

Spreading

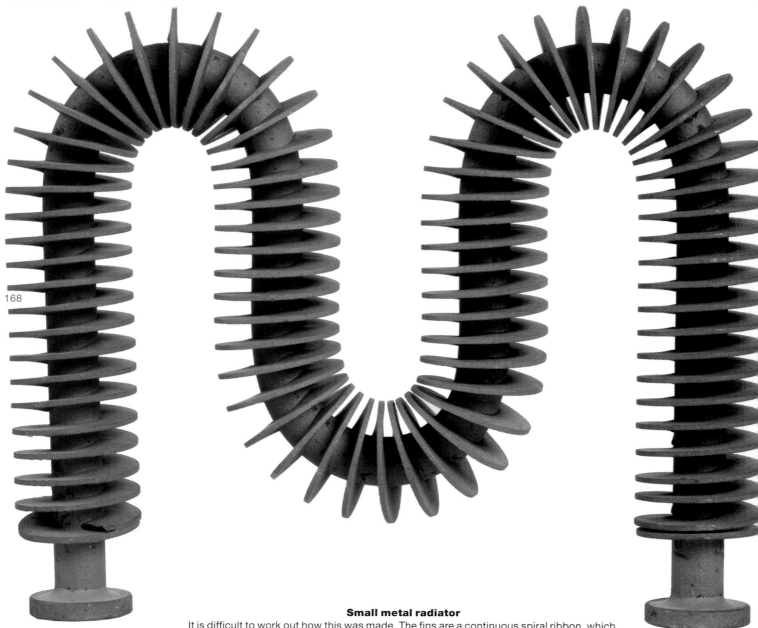

Small metal radiator
It is difficult to work out how this was made. The fins are a continuous spiral ribbon, which, given its delicacy, could not have been part of the inner tube when that was bent into shape.
18 cm/7 in.

Spreading

Water heater element
Made of copper.
15 cm/5¾ in.

Two heat sinks
Above: Secured to a piece of electronic equipment to keep it cool. A section of extruded aluminium.
11 cm/4¼ in.
Right: Fans are mounted on heat sinks to speed up heat loss. The extruded aluminium fins are presumably curved against the direction of the rotating column of air.
10 cm/4 in.

Dentist's tooth shade guide
Matching the colour of a crown or filling to existing teeth is easier if the dentist can hold a sample against its neighbours.
11 cm/4¼ in. diameter

Testing
To check, calibrate, survey, mark out, estimate, gauge, judge or measure

Adjustable drafting template
Before the advent of computers, designers used templates of metal, wood or plastic to draw small circles, ellipses and polygons. This template was useful for drawing courses of brick or tile on building plans and elevations.
15 cm/5¾ in.

If hitting is the first and most primitive activity for which our earliest ancestors developed tools, measuring and testing must be at the opposite end of the scale of technological development.

The tools in this chapter measure or test electricity, wind speed, temperature, distance, hearing, strength, angle, circumference and eyesight. They are used in trades and occupations as diverse as coach building, tailoring, sculpture, optometry and gunnery.

My interest in drawing and measuring instruments stems from a career spent using or teaching others to use them. From instruments used by designers on drawing boards, my collection expanded to include those used by carpenters, sculptors, engineers and masons on wood, plaster, metal and stone.

My eyes were opened to the beauty of compasses, callipers and dividers by a visit to the annexe of the Pompidou Centre in Paris, which houses a replica of Brancusi's studio. The tools hanging on the wall of his workshop seemed to me as sculptural as the works in the main room. That Brancusi drew inspiration from anonymous objects is suggested by the presence, shown in photographs of the original studio, of a large wooden screw thread from a wine press leaning next to early versions of *The Endless Column*.

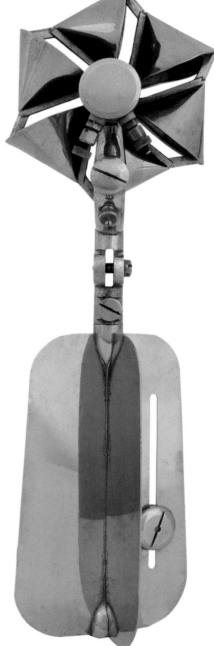

Naval anemometer
An instrument for measuring wind speed. The revolving metal cones are reminiscent of Greek windmills and of the conical sieves in Duchamp's *The Large Glass*.
37 cm/14½ in.

Testing

Voltage tester
An early device for testing electrical current.
21 cm/8¼ in.

Watch spring gauge
Watchmakers need to identify both the width and the diameter of springs.
17 cm/6¾ in.

Builder's level and plumb
The eye and the mouth of this tool allow a carpenter or bricklayer to check vertical and horizontal work.
65 cm/25½ in.

Testing

A pair of plastic-coated wrist weights
The moulded, hand-friendly shape is also found on the handles of knives, guns and ski sticks. The wrist loops have been removed.
Both 13 cm/5 in.

Cast-iron wrist weight and grip exerciser
A mini dumb-bell, sliced to incorporate springs.
16 cm/6¼ in.

Testing

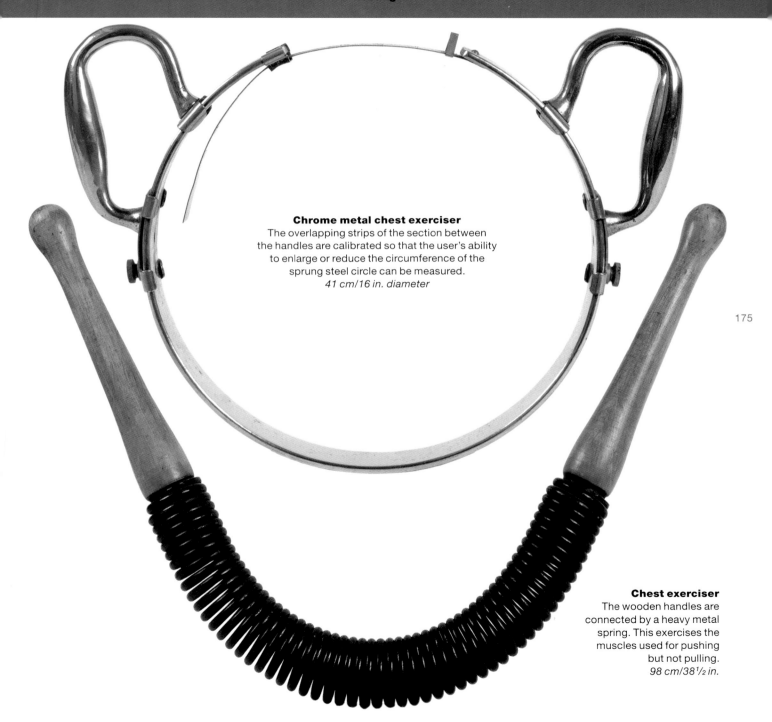

Chrome metal chest exerciser
The overlapping strips of the section between the handles are calibrated so that the user's ability to enlarge or reduce the circumference of the sprung steel circle can be measured.
41 cm/16 in. diameter

Chest exerciser
The wooden handles are connected by a heavy metal spring. This exercises the muscles used for pushing but not pulling.
98 cm/38½ in.

Hatter's gauge
This measures the internal circumference of a hat. A similar mechanism to that used by the chest exerciser on the previous page.
13 cm/5 in. diameter

Testing

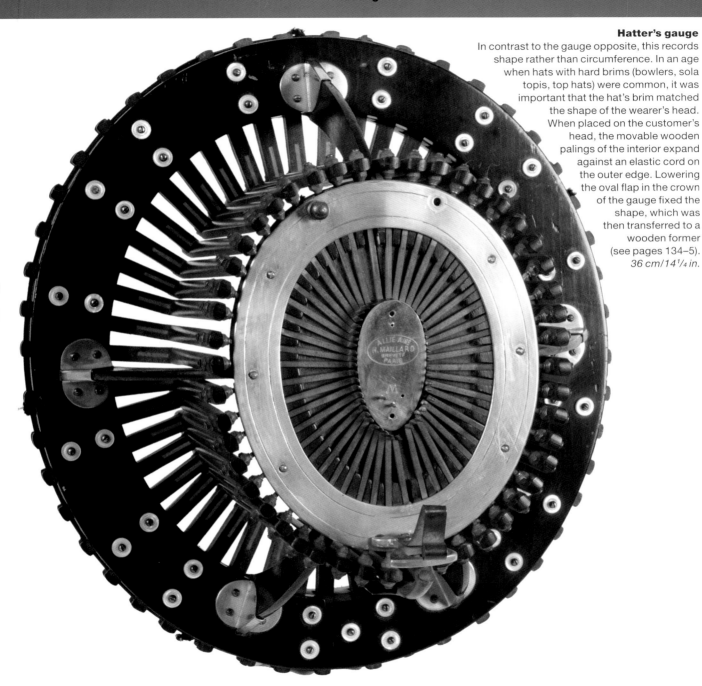

Hatter's gauge
In contrast to the gauge opposite, this records shape rather than circumference. In an age when hats with hard brims (bowlers, sola topis, top hats) were common, it was important that the hat's brim matched the shape of the wearer's head. When placed on the customer's head, the movable wooden palings of the interior expand against an elastic cord on the outer edge. Lowering the oval flap in the crown of the gauge fixed the shape, which was then transferred to a wooden former (see pages 134–5).
36 cm/14¼ in.

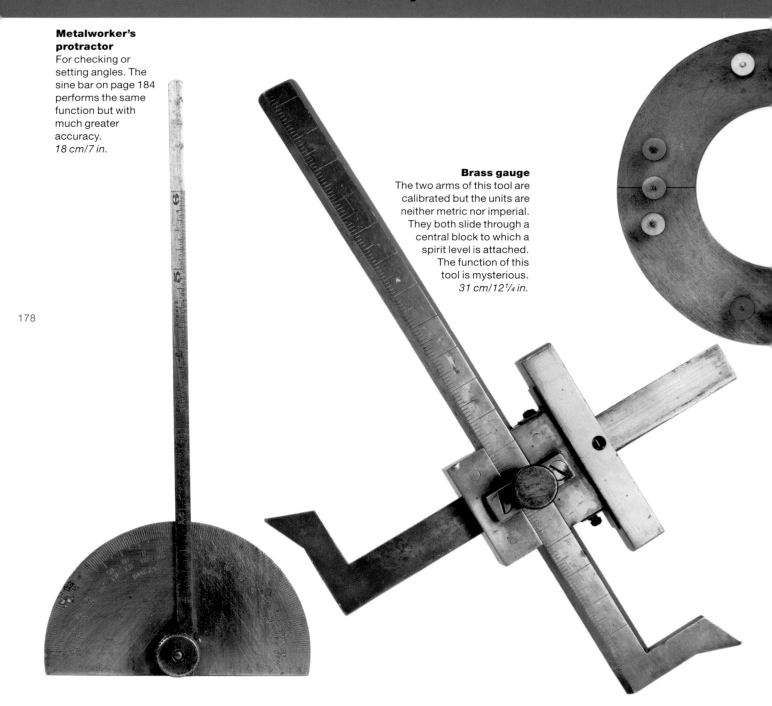

Metalworker's protractor
For checking or setting angles. The sine bar on page 184 performs the same function but with much greater accuracy.
18 cm/7 in.

Brass gauge
The two arms of this tool are calibrated but the units are neither metric nor imperial. They both slide through a central block to which a spirit level is attached. The function of this tool is mysterious.
31 cm/12¼ in.

Testing

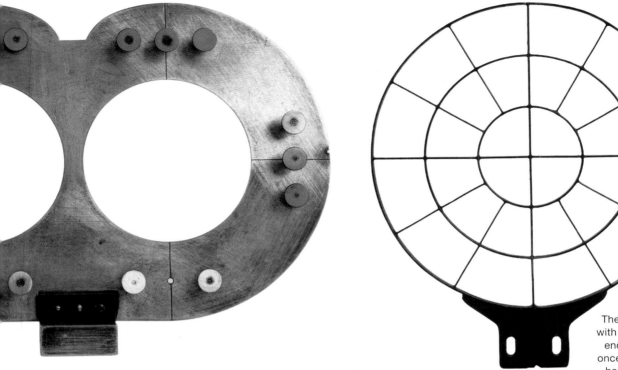

Unknown object
The threaded brass pins may have held wires that formed a grid over the two circular holes. The back face is marked in degrees, as on a protractor. Was this placed in front of a fixed pair of binoculars?
28 cm/11 in.

Anti-aircraft gun sight
The centre cross aligned with a second point at the end of the barrel, which, once the estimated range had been adjusted, was then aligned with targets in the field of fire.
22 cm/8¾ in. diameter

Cake templates
These help a sous-chef cut pies and tarts into equal segments.
Up to 18 cm/7 in. diameter

Testing

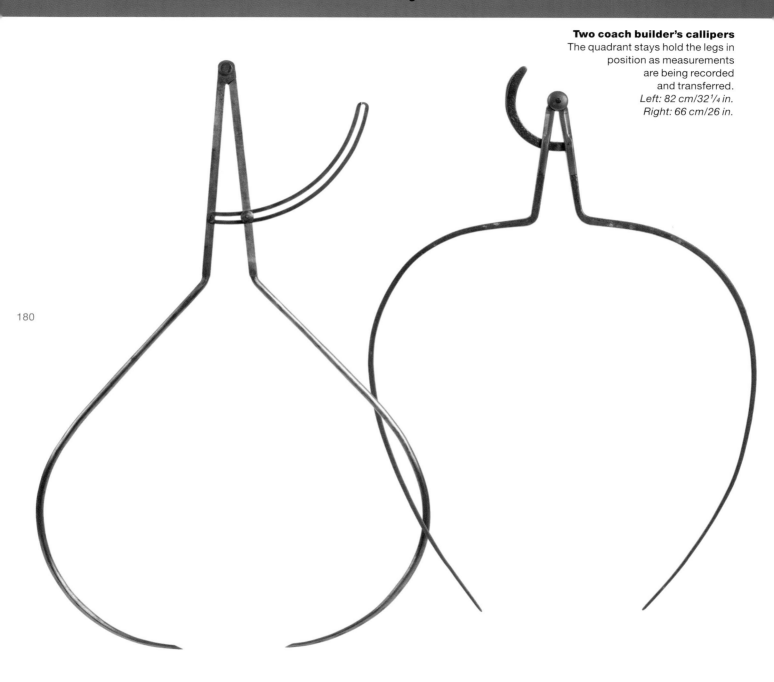

Two coach builder's callipers
The quadrant stays hold the legs in position as measurements are being recorded and transferred.
*Left: 82 cm/32¼ in.
Right: 66 cm/26 in.*

Testing

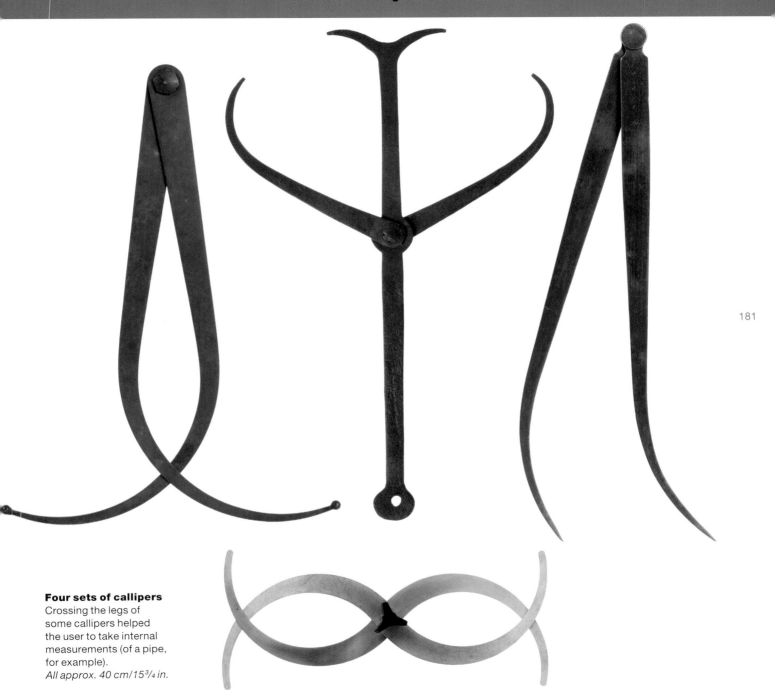

Four sets of callipers
Crossing the legs of some callipers helped the user to take internal measurements (of a pipe, for example).
All approx. 40 cm/15¾ in.

Four drawing tools
From left to right:
A stone-mason's template for drawing curves on stone prior to carving an Ionic capital.
43 cm/17 in.
A tailor's trouser rule for measuring and marking out shapes to be cut from cloth.
80 cm/31½ in.
A carpenter's trammel for marking out curves on timber. A robust alternative to the compass.
38 cm/15 in.
A Portuguese carpenter's T-square. The bar can be adjusted to any desired position on the shaft.
31 cm/12¼ in.

Testing

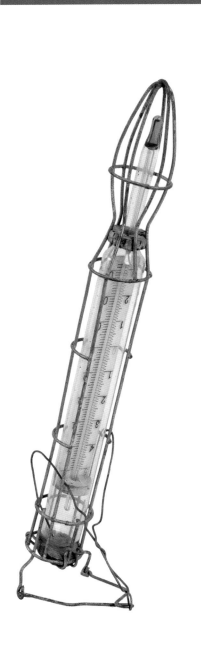

Four testing tools
From left to right:
A Portuguese winemaker's thermometer for checking the temperature of fomenting Madeira.
23 cm/9 in.
A railway engineer's track gauge for checking the distance between rails. Given the scale of the work, something stronger than a tape measure is required. This tool is 160 cm (63 in.) long and cut from 9 mm (¼ in.) steel plate. Art scholars may detect an echo of Duchamp's *3 Standard Stoppages*.
160 cm/63 in.
A doctor's tuning fork for testing hearing. After striking the tines, the heel of the fork is placed against the bone behind the ear to check the patient's sensitivity to vibration.
19 cm/7½ in.
A jeweller's ring gauge. The ring is slid down the calibrated, conical shaft to measure the internal diameter: much easier than using a micrometer. There is, presumably, a jeweller's method for measuring customers' fingers, possibly a variant of the apple gauge on page 188.
19 cm/7½ in.

Testing

Sine bar
A metalworker's tool for recording or setting out angles. If the distance between the two metal discs is known, when one end of the bar is raised to a known height the angle of the bar to the horizontal can be precisely calculated using the sine/cosine rule. This is more accurate than taking a reading from a protractor.
27 cm/10½ in.

Metalworker's tool
For checking the accuracy of circular objects. The fact that this is marked with the owner's initials rather than a manufacturer's stamp suggests that this may also be an apprentice's test-piece.
23 cm/9 in.

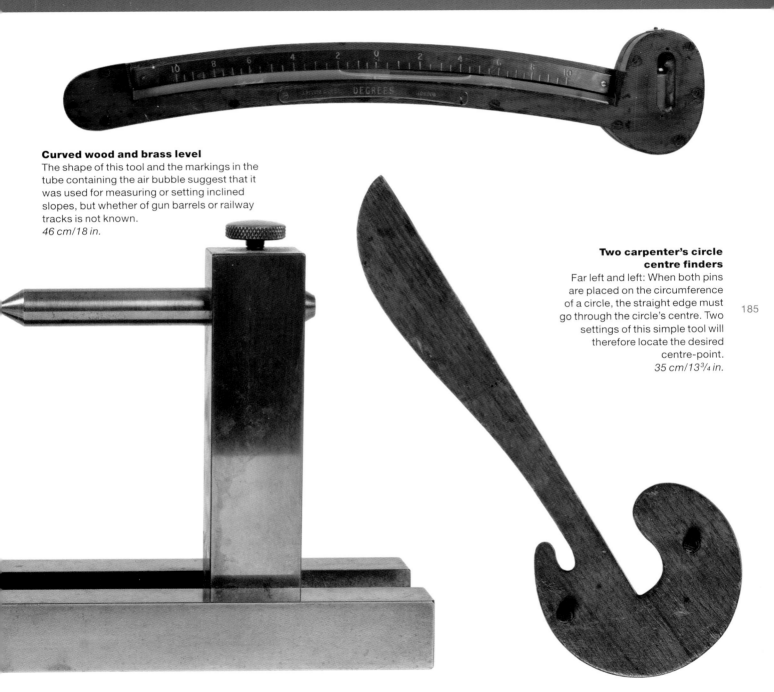

Curved wood and brass level
The shape of this tool and the markings in the tube containing the air bubble suggest that it was used for measuring or setting inclined slopes, but whether of gun barrels or railway tracks is not known.
46 cm/18 in.

Two carpenter's circle centre finders
Far left and left: When both pins are placed on the circumference of a circle, the straight edge must go through the circle's centre. Two settings of this simple tool will therefore locate the desired centre-point.
35 cm/13¾ in.

Testing

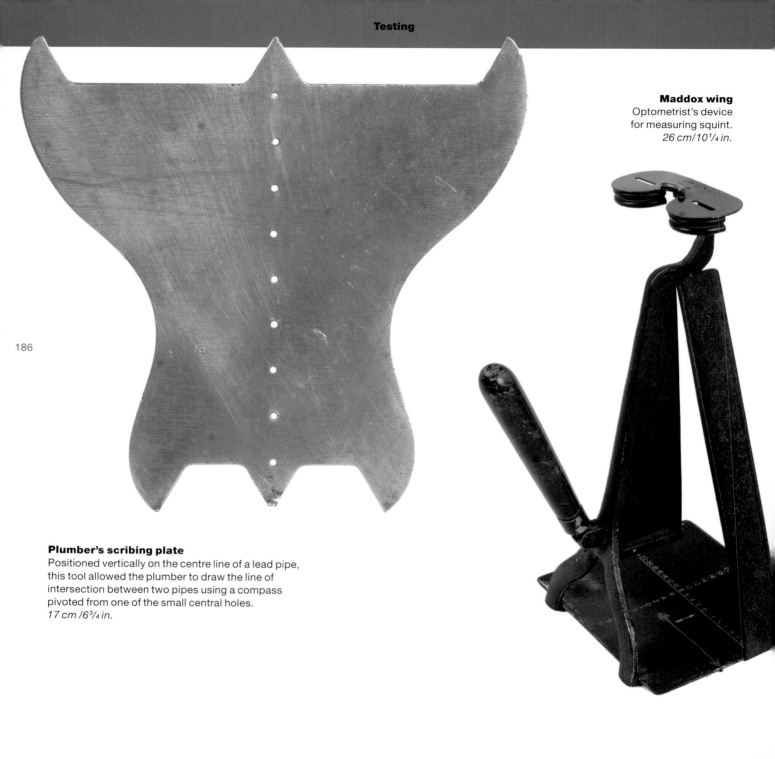

Maddox wing
Optometrist's device for measuring squint.
26 cm/10¼ in.

Plumber's scribing plate
Positioned vertically on the centre line of a lead pipe, this tool allowed the plumber to draw the line of intersection between two pipes using a compass pivoted from one of the small central holes.
17 cm /6¾ in.

Testing

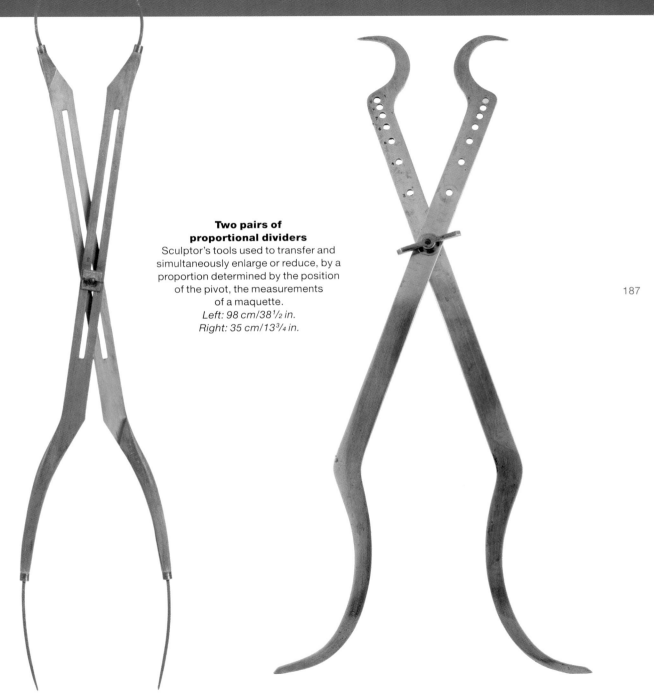

Two pairs of proportional dividers
Sculptor's tools used to transfer and simultaneously enlarge or reduce, by a proportion determined by the position of the pivot, the measurements of a maquette.
Left: 98 cm/38½ in.
Right: 35 cm/13¾ in.

Testing

Plywood apple gauge
Supermarkets are apparently fanatical in their rejection of fruit above or below a standard size.
36 cm/14¼ in. diameter

Testing

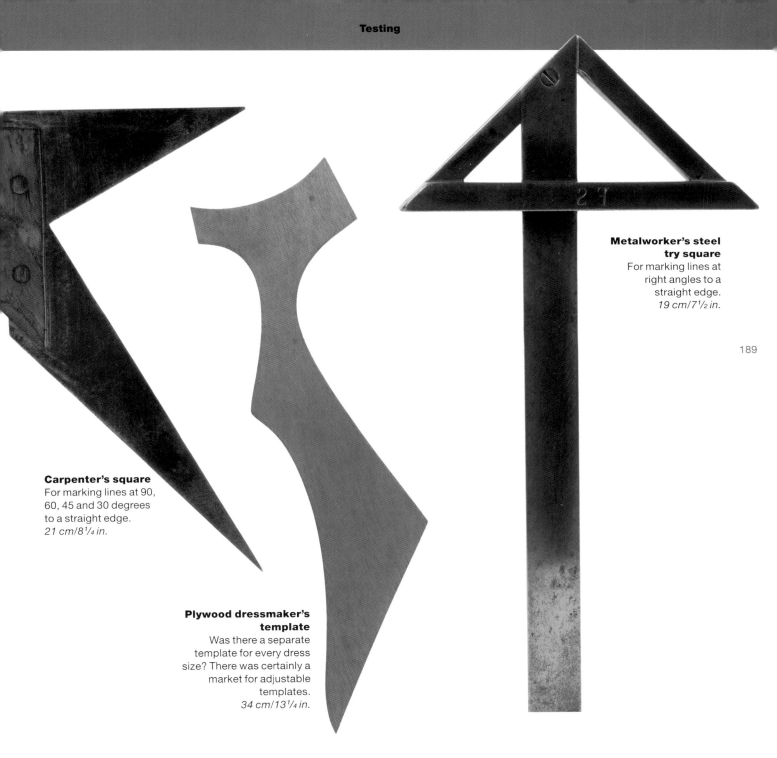

Carpenter's square
For marking lines at 90, 60, 45 and 30 degrees to a straight edge.
21 cm/8¼ in.

Plywood dressmaker's template
Was there a separate template for every dress size? There was certainly a market for adjustable templates.
34 cm/13¼ in.

Metalworker's steel try square
For marking lines at right angles to a straight edge.
19 cm/7½ in.

Optician's trial frame
Different lenses are slotted into the frame to test the patient's eyesight. A more sadistic version of this device, which held the eyelids open, features in Stanley Kubrick's film *A Clockwork Orange*.
16 cm/6¼ in.

Folding opera glasses
These dispense with the conical tube that usually separates the two lenses. Opera-going also led to the invention of the opera hat, a top hat that folded flat when sat on.
14 cm/5½ in.

Testing

Collapsible binoculars
The tube connecting the two lenses, instead of being a nest of cylinders as in a telescope, is made from a spiral ribbon of aluminium.
12 cm/4¾ in.

Index

Agitator, curd 63
Anchor 75
Anemometer 172
Anvil 28–9
Auger 48
Axe 26, 32, 56, 75
Bag, iced-water 104
Base, theodolite 2
Beater, carpet 30–1
Bed-stick 100
Bells 165
Binoculars 191
Bit, horse's 78
Blade, axe 26
Blade, knife 26, 56
Blade, food mixer 110
Blade, pickaxe 26
Blade, lawnmower 52
Blade, steel mill 60
Blade, turbine 51
Blade, waste grinder 106
Boules 90
Boot 130
Brush 162
Buttris, farrier's 62
Callipers 180–1
Carpentry tools 48–9, 81, 173, 182, 185, 189
Case, pipe 105
Cash till 86
Castrator 46
Caulking iron 26
Celt 32
Chisel 48
Chopping board 118
Clamp, dipping 73
Clamp, grinder 107
Clamp, mitre 80
Clamp, mole 14
Clamp, saddler's 78
Clamp, vine 80
Clamp, welding 96
Cleaver 33
Clothing 75, 130–1
Club 26
Coffee maker 18, 19
Comb 120
Compressor, valve spring 82
Connector, gas 10
Cornet 166
Coronet, skull 70
Cover, pipe 129
Crampon, whaler's 74
Creaser, trouser 140
Crutches 100–1
Cutter, baguette 44
Cutter, biscuit 63, 64

Cutter, chip 63
Cutter, egg 65
Cutter, pasta 45
Cutter, washer 64
Cylinder head, motorcycle 4, 158–9
Deadeye block 74
Dentistry 16, 149, 170
Depressor, tongue 79
Diaphragm 155
Dibber 34
Dilator 160
Drill 55
Element, heater 169
Exerciser 174–5
Fan 163
Feeder 104
Fly whisk 163
Football 92–3
Forceps 70–1
Fork, carving 23
Fork, lifting 83
Fork, stable 99
Fork, toasting 30–1
Fork, tuning 83
Forcer, cucumber 143
Former 143, 146, 160
Gag 79
Gardening tools 7, 34–5, 38, 43, 46, 48, 52–3, 57, 80, 82
Gauge 173, 176–8, 183, 188
Glasses 123, 190
Gouge 114
Guillotine, tonsillectomy 40, 50
Gun sight 179
Hackle board 121
Hammer 24–5, 26, 58
Hanger 94–5
Harpoon 6
Hat 130–1
Hat former 147, 160
Hat gauge 176–7
Hat mould 135
Headrest 9, 88, 96
Heat sink 169
Hoe 7, 34, 57
Holder, egg 96
Holder, roast meat 85
Holder, ski boot 83
Holder, wine bottle 102
Hook 37, 87
Horn 157, 165
Hose spreader 162
Hot-water bottle 104
Hunting equipment 6, 27, 36–7, 74, 101
Ice tools 20, 26, 74–5

Inhaler 77
Insulator 128–9
Jet, gas 164
Kitchen utensils 18–19, 23, 30–1, 33, 39, 44, 60, 85, 96, 8–9, 102, 106–7, 110–11, 113, 115, 118, 128
Knife 26, 49, 56, 60, 62
Last, shoe 147
Lawnmower 52
Lead 97
Leque 89
Level 173, 185
Lopper 43
Mask, rubber 97
Mask, hockey 122
Mask, fencing 127
Massager 112, 116–7
Masticator 51
Medical equipment 5, 6, 16, 24–5, 41, 50–1, 54–5, 68, 70–1, 76–7, 97, 100–1, 104, 116, 148–9, 160, 171, 186, 190
Milk pot 103
Mixing bit 114
Model, ear 148
Model, foot 148
Model, jaws 16, 149
Model, liver 5
Model, stomach 148
Model, stone-crushing plant 109
Motoring equipment 4, 8, 12, 66, 72, 73, 84, 158–9
Mould 132, 135, 136, 138–9, 140–1, 142, 146
Mug 76
Multi-tool 26, 58, 73
Musical equipment 89, 157, 165, 166
Muzzle 17
Needle 36
Nutcracker 58
Oil can 103
Oil filter band 72
Opener, bottle 58, 84
Palette 91
Pattern 14, 136, 140–1, 144–5
Perforator 54
Pestle 110
Pick 26
Pickaxe 26
Picker, fruit 82
Pillow, wooden 9, 88, 96
Pineapple 124
Piston ring expander 84

Planter, seed 34
Plaster body part 5, 148
Pliers 85
Pounder 111
Propeller 40
Proportional dividers 187
Prosthesis, hip 6
Protractor 178
Pruner 53
Pump 155, 162
Rabbit, taxidermist's 134
Rack 101
Radiator 168
Rake 34
Retort 103
Retractor, obstetrical 70
Retractor, surgical 71
Ring, gas 153, 164
Rip, slater's 62
Roaster, apple 85
Roller 111–13
Rongeur 54
Rotivator 34, 38
Rule 182
Saddle 90
Salad basket 98
Saw 49, 55
Scissors 59
Scraper 121
Screw 82
Screwdriver 6, 26, 58
Screw thread 12
Scribing plate 186
Seat, bicycle 90
Seat, rowing 90
Secateurs 46, 48, 53, 54
Shears 50
Shield 126
Shoe tree 146
Sieve 98
Sign, shop 167
Sine bar 184
Snow shoe 75
Snuffer, candle 85
Spanner 72–3
Spatula 114
Spear 6, 36–7
Speculum 79
Spindle 95
Splint 100
Spoon 98–9
Sporting equipment 15, 90, 92–3, 122, 127
Spring, bed 100
Spring, car window 150
Spring, gramophone 151
Square 189
Stand, cake 102

Stand, carving 102
Stand, iron 129
Stand, pipe 105
Stay, pack 78
Stay, printing press 12
Stone tools 32
Stretcher, leather 142
Stretcher, neck-tie 143
Stretcher, pipe 161
Stretcher, teat 161
Stringer, racquet 72
Stripper, cable 58
Stripper, corn cob 119
Sugar shaker 152
Swift 94
Syringe 154, 156
Tank, developing 114
Template 171, 179, 182, 188
Tenderizer, meat 39
Thermometer 183
Trammel 182
Trimmer, hoof 47
Trimmer, nail 58
Trivet 128
Truss 68
Tongs 18, 19, 20, 67, 74, 84
Toy fist 22
Trap, fish 101
T-square 182
Tuning fork 183
Tyre iron 8
Urinal 76
Veterinary equipment 79, 154, 156
Vice 80
Voltage tester 173
Weights 174
Whisk 115
Wing, Maddox 186

References

Page 8:
Goncourt Journal: from Robert Baldick, ed. and trans., *Pages from the Goncourt Journal* (The Folio Society, London, 1980).

Page 20:
Paul Nougé: from Paul Nougé, 'Les images défendues', in *Le Surréalisme au service de la révolution*, no. 5 (Paris, 1933).
Claude Cahun: from Claude Cahun, 'Prenez garde aux objets domestiques', in *Cahiers d'Art*, nos. 1–2 (Paris, 1936).
Robert Hughes: from Robert Hughes, *Things I Didn't Know: A Memoir* (Alfred A. Knopf, Inc., New York, 2006).
Georges Franju: from an interview featured on *Les Yeux sans visage* DVD edition (2004).
David Smith: from David Smith Archives, Box 27, Miscellaneous Writings Dated and Undated, Writings #4 (Undated).

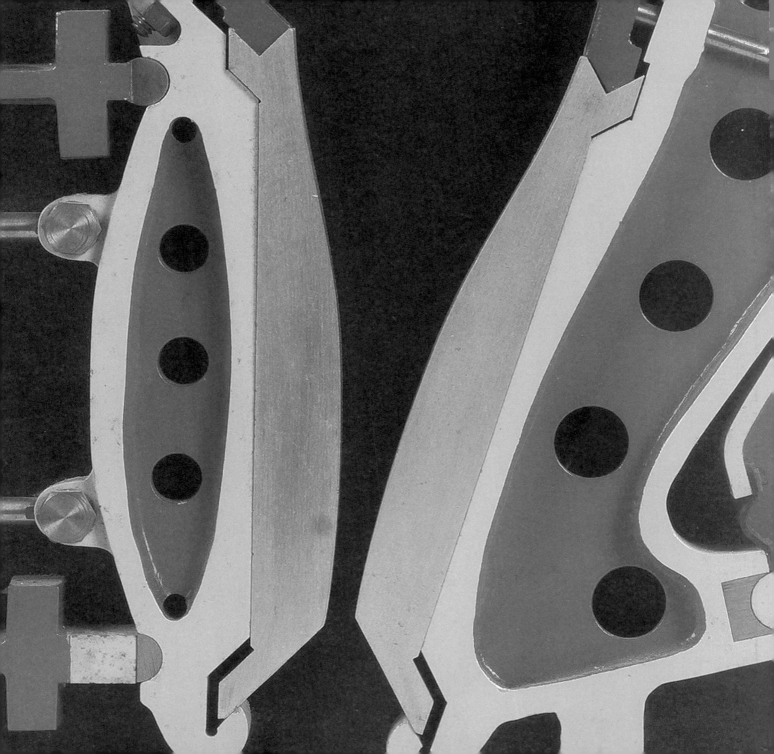